Lasers—How They Work

About the Book

The laser is a ray of "cool" concentrated energy on the move at the speed of light. It is so bright it outshines the center of the sun. Clearly and absorbingly author Charles H. Wacker, Jr., tells the fabulous story of the laser—how it is being developed, what it can already accomplish. For example, the conventional means of drilling a hole in a diamond takes three days, yet the laser can do it in ten minutes at a cost of a few cents. Wacker reveals that in the future everything in our lives will feel the impact of the laser, from a ray of light which will propel us through space to a weed killer.

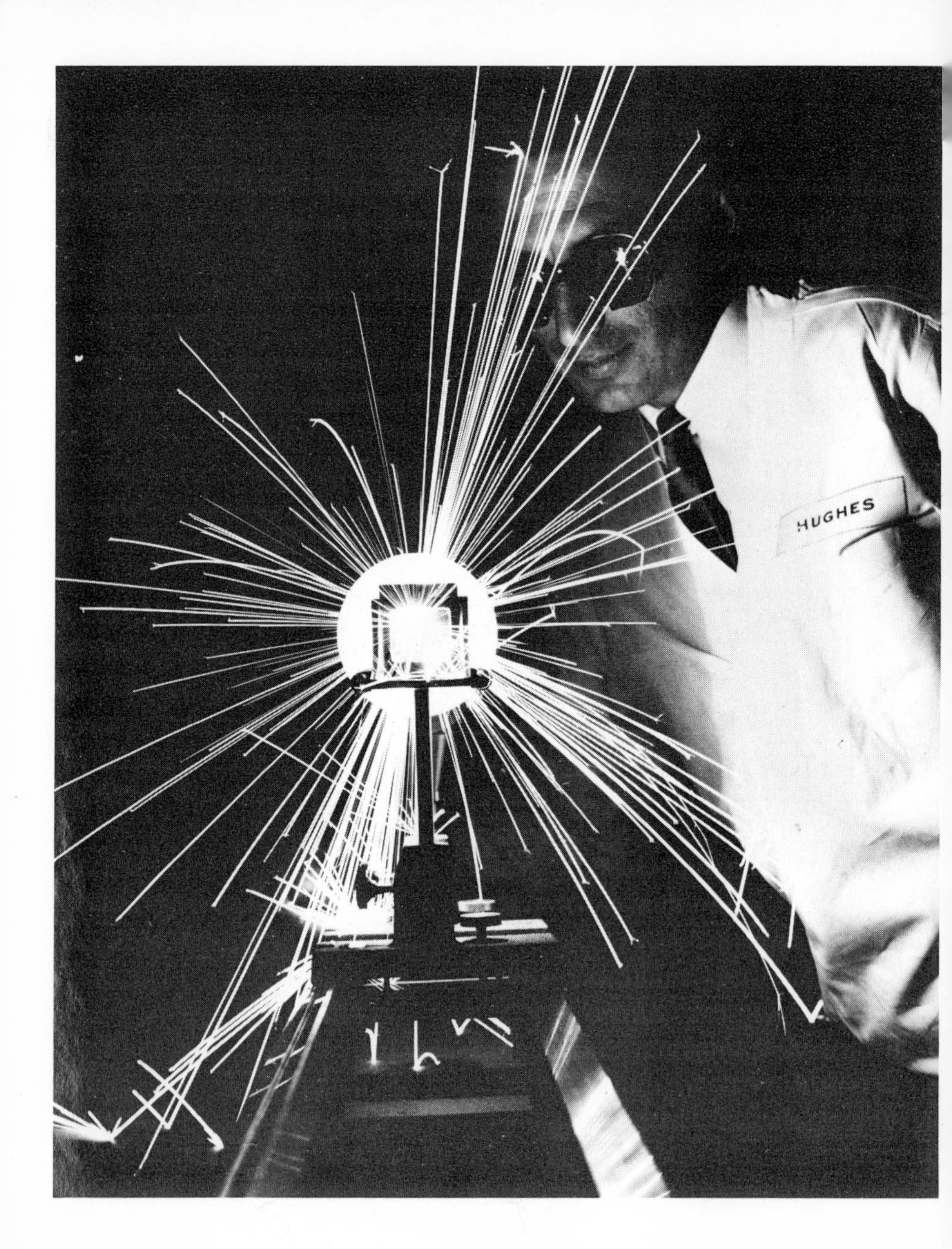

A ruby laser piercing a hole in a sheet of extremely hard tantalum metal. A Hughes laser technican observes the intense burst of the beam which requires 1/1,000 second to disintegrate metal with a boiling point of 10,000° F. (*Courtesy Hughes Aircraft Company*)

LASERS
HOW THEY WORK

By Charles H. Wacker, Jr.

G. P. Putnam's Sons • New York

SBN: GB-399-60841-9
SBN: TR-399-20352-4
Library of Congress Catalog Card Number: 79-170076
PRINTED IN THE UNITED STATES OF AMERICA
12216

Contents

Acknowledgments

I am especially grateful to Dr. Theodore H. Maiman, discoverer of the ruby laser, and to Mr. Frank Crean, Science Department chairman, Beverly Hills, California, High School, who critically read the original manuscript and made many valuable suggestions which are incorporated in the finished work.

Foreword

Man long has been fascinated by the idea of a knife so sharp it might perform delicate operations without cutting the skin. Conversely, he has been just as intrigued by the death rays of early science fiction. And he has looked forward to the day when he would be able to shine a beam of light into the darkness and project his voice, his image, himself upon it. Such radiant energy is powerful and pure, and scientists had long anticipated what we now call the laser.

In the seventeenth century Isaac Newton said that light was made of corpuscles emitted by luminous bodies and that light was made of different colors which in combination produced white light.

Although the Newtonian corpuscular theory was soon superseded by other theories, such as Christian Huygens' wave theory of light, the problem of defining and explaining light continued to challenge scientists.

A. A. Michelson and E. W. Morley, collaborating British scientists, used two beams of light to discover the ether wind created by the earth rushing on its journey around the sun.

James Clerk Maxwell, the Scottish mathematician, studying previous experiments with electricity and magnetism, produced his *Treatise on Electricity and Magnetism,* which contained key formulas on light and predicted electromagnetic waves.

Michael Faraday, with his magnetic lines of force experiments, further equated light and magnetism. And the German scientist Heinrich Hertz conducted a simple experiment (a space, or two spark gaps, across which lightninglike bolts of electricity would leap) to demonstrate Maxwell's formulas on light and magnetism.

However, it was not until the twentieth century that the technology was available to provide researchers with a true light tool in the form of the laser.

Although the laser holds great promise, many potential applications must still await further developments in circuits and components before its ultimate value to mankind can be explored.

Lasers—How They Work

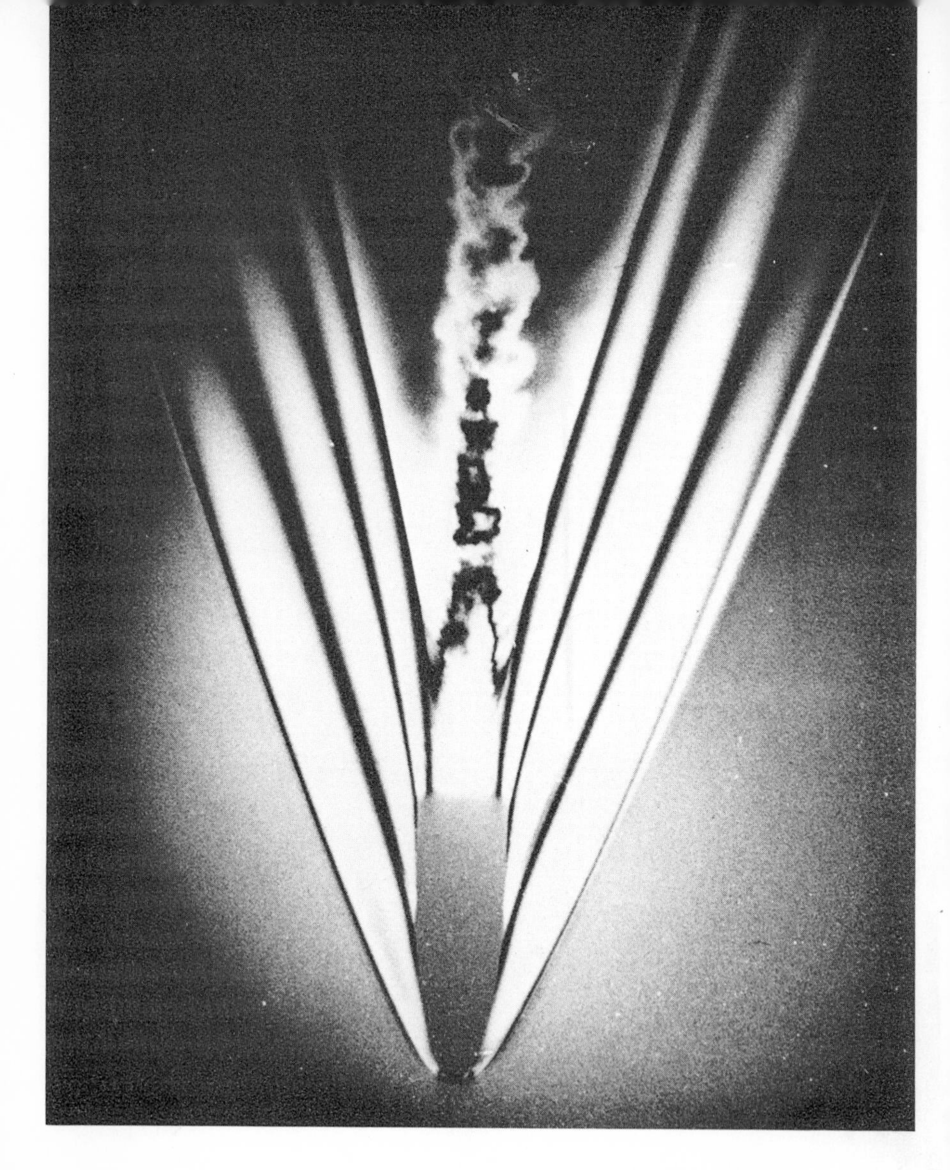

A double-exposure holographic interferogram of a .22-caliber bullet traveling at 3,500 feet per second showing the shock waves surrounding it, taken with a fifty-billionths-of-a-second ruby laser by TRW Systems' scientists. (*Courtesy TRW Systems Group, TRW*)

1
Definitions and Developments

The laser is a new experience in light. It is a ray of "cool" concentrated energy on the move—it is light so bright that it outshines the center of the sun. It is atomic light. The rapid movement of its atoms creates tremendous power in the form of electromagnetic energy which can instantaneously pierce steel.

A laser beam can travel vast distances at the fabulous speed of 186,000 miles per second. It can pinpoint a target hundreds of thousands of miles away in the dark reaches of space. The best comparison to laser radiation is sunlight. It is energy at work. Focus sun rays on a leaf with a magnifying glass and they will burn a hole in it. Let the rays fall on plants and they will grow. Focus laser beams through a prism and they will combine so much energy at the point of focus (nearly two times the surface temperature of the sun–32,000° F) that they will literally vaporize the area of focus without so much as singeing the rest of the substance.

The word “laser” is an acronym of the first letters of the five main words which describe its function: Light Amplification by Stimulated Emission of Radiation.

The laser is like an electronic oscillator, an amplifier with feedback. It amplifies light to a very high intensity using an oscillating medium, an external source of energy, and a resonant circuit with sufficient feedback to sustain oscillation.

The oscillator is the “photon” action. The energy source may be either electrical or optical. The resonant circuit consists of the photon source (the atomic material) and a pair of adjusted reflecting surfaces at either end of the laser medium.

Ordinary, incoherent, “noisy” light, such as we see all around us, would burn itself up in achieving laser energy. Laser light is coherent and cool. Its energy travels in rhythm and its effective temperature is in billions of degrees. If we were to use an ordinary light source to try to equal the energy of the laser, it would disintegrate in the process because of the heat generated. Most of the light sources we see are hot. The filament actually burns to produce energy. No matter how thick the filament and the globe surrounding it, they could not withstand the temperature required by the laser.

The laser was something that industry wanted for a long time. It had been under development in laboratories in the United States, England, Japan, and Russia since about 1958. Most of the research was done in America, and this by a few large electronic firms.

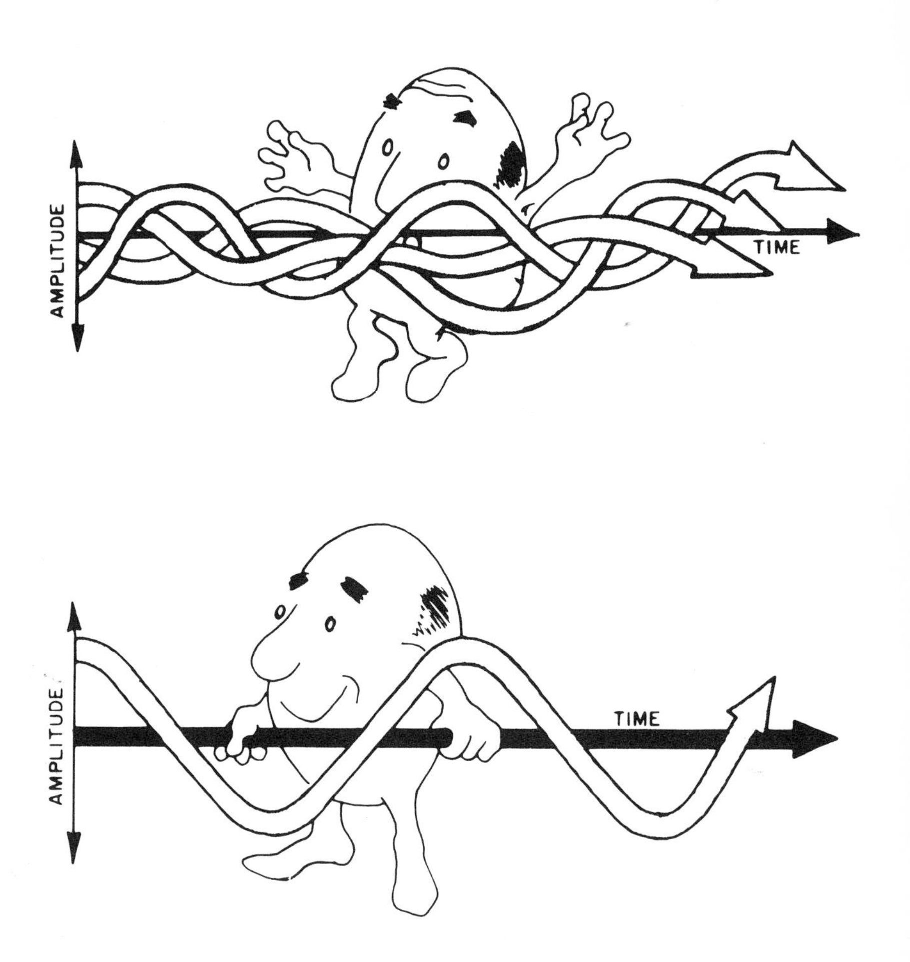

(Top) Incoherent light — Ordinary light from an incandescent bulb is made up of many radiations in the photoelectric spectrum. It is random, consisting of many wave fronts which interfere with each other and cause cancellation.
(Bottom) Coherent light — Laser light is coherent. It is of a given wavelength, directional, with its wave fronts reinforcing each other. This coherence is responsible for the enormous concentrated energy present in a laser beam. (Courtesy RCA)

It became a reality on July 7, 1960, having been announced at a press conference in New York City by the man who invented the first successful "ruby" laser, Dr. Theodore H. Maiman. He had made his discovery while working as a senior scientist at the Hughes Aircraft Company, Culver City, California.

The magic in the word "laser" gave its announcement credibility, despite the fact that its scientific principles were too complex for public understanding.

Unbounded optimism by scientists and laymen caused all sorts of predictions to be made for the "light fantastic," as it was called in those early days.

As with most major scientific discoveries, individuals had had the laser idea almost since the start of the twentieth century. In 1900 Max Planck, a German physicist, developed the quantum theory, which explained how energy is produced by atoms and molecules in constant "bundles" or "quanta." Energy under certain conditions is radiated in tiny light balls and not always as waves, as conceived by Huygens, Augustin Fresnel, and Thomas Young.

Dr. Albert Einstein further opened the laser door by combining the quantum theory with his photoelectric concept. The quantum became the "photon," a subatomic unit that stimulated energy emission.

Scientists had to solve the duality problem of waves and particles in order to cross the laser threshold. Ernest Rutherford, an English physicist, in probing the riddle laid the foundation that enabled Niels Bohr, the Danish scientist, to solve it. He conceived of atoms as a nucleus surrounded by electron levels. While the electrons remain on any of these levels they are nonenergy-producing. Each level represents a certain degree of energy: the higher the level, the greater the energy potential of the electron. Electrons are capable of changing levels, but in order to do so they need a boost—they must be stimulated by an external power source such as light or current to "move upstairs." An electron that has been boosted to an upper energy level is "excited," while one which is on the lower or ground level is at rest.

According to Bohr's theory, in all matter there is some traffic between energy levels but the preponderance of electrons is at the ground level resting. In order to make light it is necessary to move the majority of atoms upstairs. This can be done by satiating the atomic material with energy. When the atoms drop back to the rest state, they spontaneously emit a sufficient number of photons to produce light.

Bohr stopped his theory at spontaneous emission. He assumed that the atoms would always randomly select their energy emission time and that there was no cooperation among the atoms. If this had been as far as scientists had gone, the laser would never have been discovered.

The significant point, which Bohr had missed, was described by Einstein in a paper published in 1916. Spontaneous emission only partially explained photon energy. Although most atoms tumbled from the energy level to the rest level at their discretion, some atoms were knocked down and not allowed this prerogative. They were responding to "stimulated emission," which is the essence of the laser.

Until the end of World War II, scientists and engineers had relied on electron tubes, such as once were in radios, to produce coherent beams of energy. The tubes served as oscillators to make steady carrier frequencies and as amplifiers to make the beam powerful enough to traverse long distances. Although electron tube oscillators can generate coherent waves in the radio-frequency region, they cannot be built small enough to carry energy beyond the lower end of the microwave region of the spectrum. The higher the frequency, the smaller the wavelength and the corresponding decrease required in the size of the oscillator. The highest frequency that could be attained with an electron tube was approximately one ten-thousandth of the frequency of the laser.

Obviously the electron tube method was impractical and would never enable scientists to achieve laser frequencies. A whole new approach to the laser threshold was needed. There would have to be a back door or perhaps an undiscovered secret passage to the upper frequency regions of the laser. The answer was found in the atoms and molecules.

In two years, 1954 and 1955, Drs. James P. Gordon, Herbert J. Zeiger, and Charles H. Townes successfully stimulated ammonia gas molecules to produce a steady stream of laserlike energy. Although the frequency level was below the visual, in the microwave, the rays were coherent, each photon—the stimulation one and the released one—was in step, and the majority of molecules were maintained in the upper energy level.

The three scientists called their device a maser, *m* signifying microwave frequency. It took into account the high- and low-energy molecules in the gas and the tendency for most of the molecules to remain at low level. The device consisted of these sections: (1) a cylinder containing electrically charged rods and into which the ammonia gas was sprayed, (2) a separator for differentiating the two energy-level molecules, (3) a resonant cavity where the photons would bounce back and forth off the walls until they oscillated at 24,000 megacycles per second and collided with other high-energy photons, and (4) a wave guide to withdraw this energy from the chamber and emit it as a steady beam of microwaves.

The ammonia gas laser was not particularly suited to the communication application for which it was intended because of its inability to be tuned to other frequencies and its very narrow band width. Additionally, the power output of the maser was too weak to be practical as an oscillator for transmitting coherent beams over extended distances. This was due to the fact that gas molecules, unlike molecules in solid materials, are too dispersed and the production of power depends on how tightly the molecules are packed together.

The ammonia maser was used as an atomic clock, since

each wave produced by this maser is like the swing of a pendulum, oscillating at 24 billion times a second. The maser can do this timekeeping in a most accurate manner, losing an estimated one second in ten thousand years.

Dr. Townes' maser achievement broke the last barrier to laser technology, and for this he was awarded the Nobel Prize in physics in 1964, along with two Russian physicists, A. M. Prokhorov and N. G. Basov, both of whom had also made significant contributions to the state of laser art.

Yet there was still the missing element that had to be found in the atomic structure which would enable scientists to build a maser in the optical region of the spectrum. Again, Dr. Townes was the prime mover, this time with his brother-in-law Dr. Arthur Schawlow.

Together they published a technical paper in 1958 that laid the groundwork for building a laser, although they anachronistically called it an optical maser—the maser, being a microwave device, would never be capable of operating at optical frequencies.

Drs. Townes and Schawlow defined the conditions under which laser energy could be produced. They realized that a new method of energy boosting was necessary, and in looking around they found the answer in a device that had been in existence in laboratories since the nineteenth century—the Fabry-Perot interferometer. The mirror chamber in this device was used to help scientists record optical interference patterns.

What Townes and Schawlow did was project this capability to the optical maser and point out that this mirror cavity should be able to boost the gas molecules by producing the required photon reflections at very high frequencies in order to make coherent amplified light.

On the basis of the famous Townes and Schawlow technical paper "Infrared and Optical Masers," scientists were spurred to greater competitive effort. The first solid state maser was built using silicon crystals and phosphorus. These were "paramagnetic" substances, and scientists were particularly attracted to them because of a certain imbalance in their atomic structure. In most substances the atoms are nonmagnetic—that is, they are paired off so their magnetic affect cancels. But in the paramagnetic materials this cancellation is not complete. Some of the atoms are missing electrons, and there is an imbalance—the material is magnetic. When these unpaired electrons are placed in a magnetic field, they can spin in one of two directions, they can line up with the poles of the external magnet, or they can oppose them. If the electrons line up with the outside poles, the electrons are said to be in a higher energy state. To get maser action from a paramagnetic material it is necessary to boost the majority of electrons to the upper state—a condition contrary to their nature but easily attainable by increasing the strength of the external magnet.

To compensate for the short operating bursts and time lag between electron shifts from the lower to the upper energy state in the two-level paramagnetic maser, a three-level paramagnetic ruby laser was developed by scientists of the Bell Telephone Laboratories. Essentially this consisted of atoms with two unpaired electrons and thus made available three energy levels. Now, with the energy gap between the top and bottom levels decreased, and with the addition of an intermediate energy-producing level, pumping and amplification could go on simultaneously.

The ruby seemed to hold the key to the laser, and discovering it was Dr. Theodore H. Maiman's achievement. The ruby is one of the amazing crystals whose atoms can be stimulated at two frequency levels—one very high in the optical region and one in the microwave region. Whereas the ruby maser performed at the lower level, Dr. Maiman made the ruby take a giant step on the electromagnetic spectrum from 100,000 megacycles per second to the vicinity of 500 trillion cycles per second.

The first laser was built in 1960 by Dr. Maiman and consisted of a man-made pink ruby crystal. This was contrary to popular scientific thought at the time which, based on Drs. Townes and Schawlow's technical paper, advocated gas as the only way to go in lasers. However, Dr. Maiman doubted that the gas laser would ever be practical enough since it required such tremendous pumping energy to keep the photons flowing. Additionally, he suspected the hidden high spectrum level of the ruby, which no one else had taken into account. Also despite current consensus, he believed that the ruby atoms could be stimulated by an outside light source without too much difficulty.

One of the main reasons why scientists had discounted the ruby was that previous calculations indicated that it had a great deal of energy which was lost. This so-called lost energy was not unattainable at all, according to Dr. Maiman's calculations; what was wrong was the mathematics of the past and not the ruby material.

Another stumbling block in the way of the ruby was a technical paper by Dr. Schawlow that emphasized that in order for solid materials to lase, at least one-half of the atoms would have to be at the energizing level at all times. This obstacle discouraged most scientists from considering the ruby as laser material.

Dr. Maiman filled his synthetic ruby with the brightest pulsed light he could generate. He polished the ends of the crystal and covered them with a brilliant reflecting material so that the photons would really resonate in the laser cavity. He wanted to be sure that the photons would have every opportunity to bounce back and forth as many times as possible to stimulate the release of the same kind of photons by knocking the electrons on the energized level to the rest level. In effect, he excited so much activity inside the laser material that the majority of the atoms was always in the high-energy state, a prerequisite for laser action. Yet there was no "red spot" on the wall indicating that the laser was operating—red because this was the frequency of the ruby and not because of its red crystalline material.

Obviously, in Dr. Maiman's mind, with all the other parameters carefully verified, the only remaining possibility for failure to lase was a flaw in the ruby material. This was the case, for when the original ruby was replaced with another one, there it was, burning bright as an "atomic radio light," according to Dr. Maiman, straight as an arrow, the beam of the world's first successful laser.

As Dr. Maiman recalled the events of 1960, his work was doubted by many scientists. Just before his breakthrough with the ruby crystal, several leading scientists advised him to give up what appeared to them a futile search. Even after the press conference announcing the laser, there were those who still questioned its validity. On the other hand, many individuals who doubted Dr. Maiman's work previously and had filed patent applications on laser ideas of their own rushed to the Patent Office to add the ruby to their list of possible laser materials.

The red ruby light on the laboratory wall of the Hughes Research Center in Malibu, California, where Dr. Maiman worked, signaled the start of a scientific gold rush. Within a few months of Dr. Maiman's discovery laboratories all over the country were uncovering and experimenting with new laser materials.

In December, 1960, Peter P. Sorokin and M. J. Stevenson of IBM used crystals of calcium fluoride doped with uranium to make the first infrared laser. It emitted coherent invisible infrared light at a wavelength of 24,000 angstroms (Å). The device did not require any special coolant, as the ruby laser did, and it could be operated conveniently at room temperatures. Because of the continued excitation of the atoms between light pulses from the outside power source, this laser required less pumping to produce photons than did the ruby.

The first solid state continuous wave laser was built by scientists at Bell Laboratories in the fall of 1961. Until then the concept of a continuous wave (CW) laser had seemed impractical because of the necessary time lapse between light pulses from the flash lamp. During this lull, the atomic movement inside the laser material stopped, or at least considerably slowed down. Ali Javan, William R. Bennett, Jr., and Donald R. Harriott, the Bell Laboratories' scientists, solved this problem by replacing the customary flash lamp with a steady mercury lamp to provide continual pumping of the atoms of calcium tungstate doped with neodymium. However, because of the extreme heat generated, the laser had to be supercooled to keep it from exploding. The power output of this particular model was very weak, amounting to a few milliwatts. But it was another key achievement in laser technology that is bearing fruit today.

Between July and November, 1962, just about two years to the day following the discovery of the ruby laser, the injection or semiconductor laser started operating. This major milestone was reached almost simultaneously by independent scientific teams from General Electric Company, Lincoln Laboratories of MIT, and IBM.

The laser was given the name "semiconductor" because that was the kind of material it used—tiny diodes made from gallium arsenide. The "injection" terminology was derived from the fact that the atoms were injected with current which stimulated the photons for emission. The advantages of the semiconductor laser are its small physical size—it can be held in the hand like a pen—and the emission of an intense narrow beam of coherent infrared energy about the width of a pinhead which makes it ideal for surgery. Because of this latter advantage, the injection laser got more publicity than the original laser. People wanted to know all about "its magical powers to heal."

There appeared to be magical powers, indeed, in a device that could use a piece of crystal, a bottle of gas, or a diode to radiate energy hotter than the center of the sun. Perhaps that is why the sun itself was sought as a source of laser power by RCA scientists in 1962. They developed a method using a curved mirror of focusing sunlight on calcium fluoride crystal containing dysprosium to produce the laser action. This sun-pumped laser gave off its radiation in the infrared region and offered great possibilities for space applications, since the customary bulky power source was eliminated.

"Chelate" is the name given to certain materials that have a cyclic atomic structure. Because of this atomic arrangement, scientists thought for a while that chelates would make good laser material. Additionally, chelates could be liquefied readily and thus easily handled in

Optically Pumped

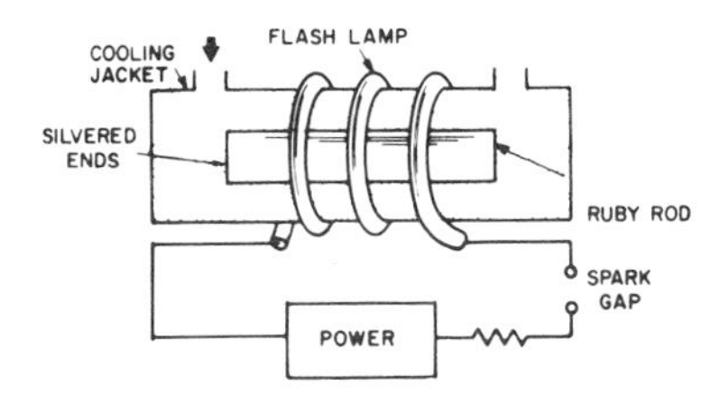

Injection

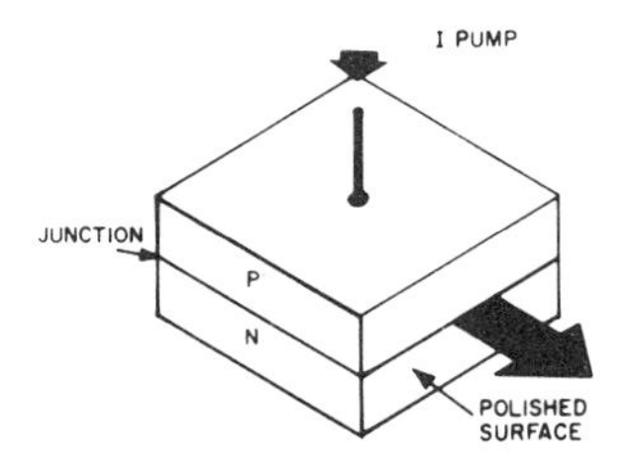

Gas

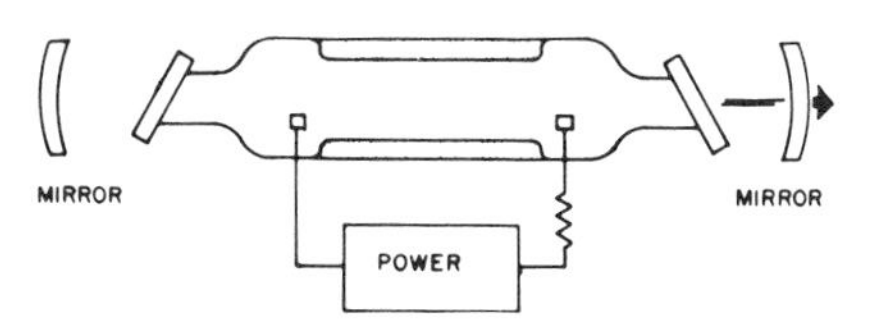

Basic laser types — Optically pumped lasers include all liquid media and solids, such as the ruby. The exception is the semiconductor type (injection). Optical pumping is usually by means of a flash lamp. Electrically pumped lasers include semiconductor and gas types and have the current fed directly into the laser material. (*Courtesy RCA*)

containers which could be made any size to meet a specific laser application. In 1963 General Telephone and Electronics Corporation scientists successfully operated a chelate laser. But the chelates were unstable during the pressures generated in the laser process.

The year 1963 also saw the introduction of the plastic laser. These plastic fibers, a little thicker than human hairs, were strung together inside glass tubes and doped with certain of the rare earths, such as ytterbium, neodymium, dysprosium, and thulium.

During 1964 several significant improvements were made in the various types of lasers. Researchers at IBM developed an inexpensive way to improve the efficiency and power output of the ruby. Simultaneously, the neodymium-doped glass laser became a strong contender with the ruby by being able to operate as a giant-pulse laser. The solid state laser attained equal footing with the gas laser in its continuous wave performance. The first high-powered gas laser was built and successfully operated using one of the gases of the atmosphere—argon. E. I. Gordon and Alan White of the Bell Telephone Laboratories designed the first "mini" gas laser using helium and neon. This combination of gases gave a very coherent, strong beam, which, unlike other gas laser beams, did not have to be trimmed continuously to stay within frequency.

Raman active substances attracted scientists for the ability to generate "natural" frequencies—that is, a frequency equal to the frequency difference between the absorbed and reemitted photons. Eric J. Woodbury and Won K. Ng, two Hughes Aircraft Company scientists, developed the first Raman shift laser in 1964. They did this by passing a ruby laser beam through a solution of nitrobenzene. The result was a wavelength shift of about 10 percent of the merging light from the characteristic ruby laser wavelength of 6,943 Å to 7,660 Å. The two beams were perfectly coherent, and this accidental discovery opened a means for scientists to study many previously unexplored wavelengths.

By 1965 laser pioneering began to level off. The principle types of lasers as we know them today had been discovered: crystal, gas, semiconductor, and liquid. Some worked well, others were questionable. During the next five years all types were either improved, changed, or rejected on the basis of new technology and need.

Perhaps it was the fascination of scientists with gaseous materials and a desire to change the seemingly fickle nature of the gas laser to be hopelessly instable which, in 1965, made C. K. N. Patel of the Bell Telephone Laboratories design a molecular gas laser. Unlike the previous atomic and ionic gas lasers with their very poor efficiency, the molecular laser increased efficiency forty times over these lasers. Now, for the first time, the extra special coherency of the gas laser, unique among lasers, became practical to apply to crucial scientific and industrial tasks.

The rough work was done: the clearing away of the stubborn obstacles of finding the right laser materials and the hewing of a reliable path through theory to operational laser. Now it was time to build on the solid foundations laid, to use the knowledge gained to make lasers more efficient, smaller, more powerful, and versatile.

To appreciate the significance of these achievements in the state of the art, we will follow in the footsteps of the laser pioneers through laser theory. Fortunately, unlike these pathfinders, we will not waste time struggling through tangles of conflicting ideas or race down laboratory trails which suddenly dead-end. Although the path through laser theory may appear difficult at times, we have simplified it and here and there have taken liberties to smooth out the complex ruts without changing the meaning of the basic ideas and concepts.

2
Laser Theory

Light is energy in motion, radiating, at work, performing a task. Conventional light sends out its rays randomly and has as its task illumination or possibly heating. Laser light, on the other hand, is coherent, its energy is concentrated in a powerful beam, and its task potential appears limitless, from communication to surgery.

All light is rated in watts, which is the power-output rating. But this power rating alone only explains one-half the ability of light to do a job. Time must also be computed to determine the work capability—that is, the length of time the energy endures. Time and energy together give the true work potential of light, and these factors are called a joule. The total number of joules in a laser beam is equal to the number of watts in the beam multiplied by the duration of the energy burst. This is the productive period of the laser.

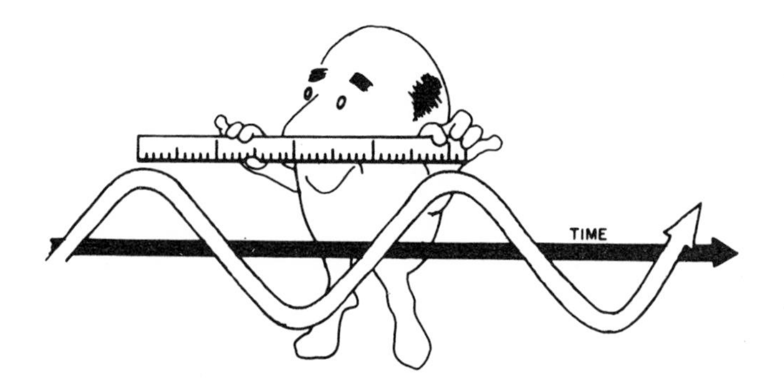

The wavelength of laser radiation may be measured in microns or angstroms (Å). (*Courtesy RCA*)

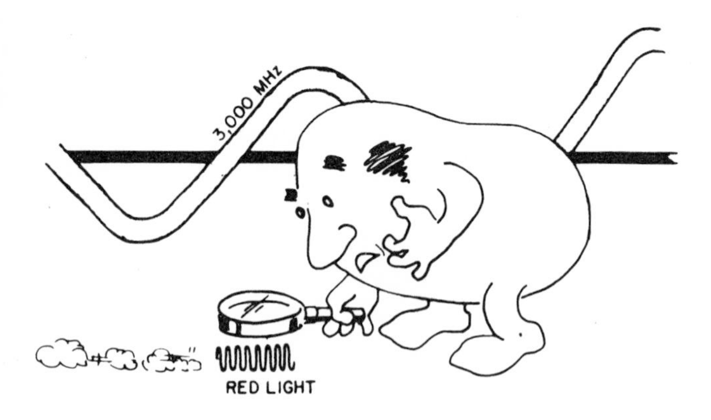

Compared to other wavelengths, such as those of radio, light waves are very short. Here, UHF at 3,000 megahertz versus the red light of a helium-neon laser. (*Courtesy RCA*)

For example, Dr. Maiman's ruby laser had an output of 10,000 watts for such a short time period that it was rated at 1/10 joule. By contrast, some large lasers have an output rating of 500 joules or more.

Laser light is also very pure. It is probably the purest light known to man. This purity is the result of the laser's light coming from a single color hue—it is monochromatic. Ordinary light, even the purest, is polychromatic, a blending of many shades. Some laser light has shone with a purity a million times greater than the color we are accustomed to seeing.

There are three basic means of identifying light: (1) velocity, (2) wavelength, and (3) frequency. A simple formula devised by scientists for finding any one of these identifiers is: Frequency x Wavelength = Velocity.

An electromagnetic wave has two parts—an electrical field and a magnetic field. These fields, or areas in which the electrical and magnetic forces are exerted, travel together. Electromagnetic waves move transversely and longitudinally. They travel at right angles to their motion. The waves move forward and the electrical and magnetic forces in them move up and down like the waves of the ocean.

Light waves are measured in millionths of an inch (angstroms, Å). The length of a wave is the distance from crest to crest or trough to trough of two wave cycles. Compared to other wavelengths, such as radio, light waves are very short. For example, the wavelength of the red light of a helium-neon laser is 63.3 millionths of a centimeter (6,328 Å), compared to a UHF radio wavelength at 3,000 megahertz of 10 centimeters.

Some lasers provide only one wavelength, whereas others may output several. A laser providing many wavelengths, or lines on the spectrum, is referred to as a multiline laser. These lasers may be operated with a prism to provide each line separately, or they may be made to perform at all lines simultaneously without the prism.

Unless great care is taken, lasers tend to output in a large number of longitudinal and transverse modes at the same time. This is called multimode operation. For example, in holography and other interference information-processing techniques, fundamental or simplest transverse mode operation is required and the laser must be carefully adjusted to meet this exacting requirement.

The same precision in the longitudinal mode is also necessary for holography. Here the coherence length (the distance over which the beam remains strongly coherent) must be maintained by minimizing the number and amplitude of the simultaneous lesser modes or "side bands" usually produced by the random motion of the emitting atoms in the laser material. One means of removing the side-band lines is by placing a special filter within the laser cavity, thereby tapering the undesirable lines. Another method is to shorten the cavity length, since this reduces the gain and limits the number of modes which can oscillate. However, there is a limit to the shortening due to the fact that this procedure reduces the laser's power.

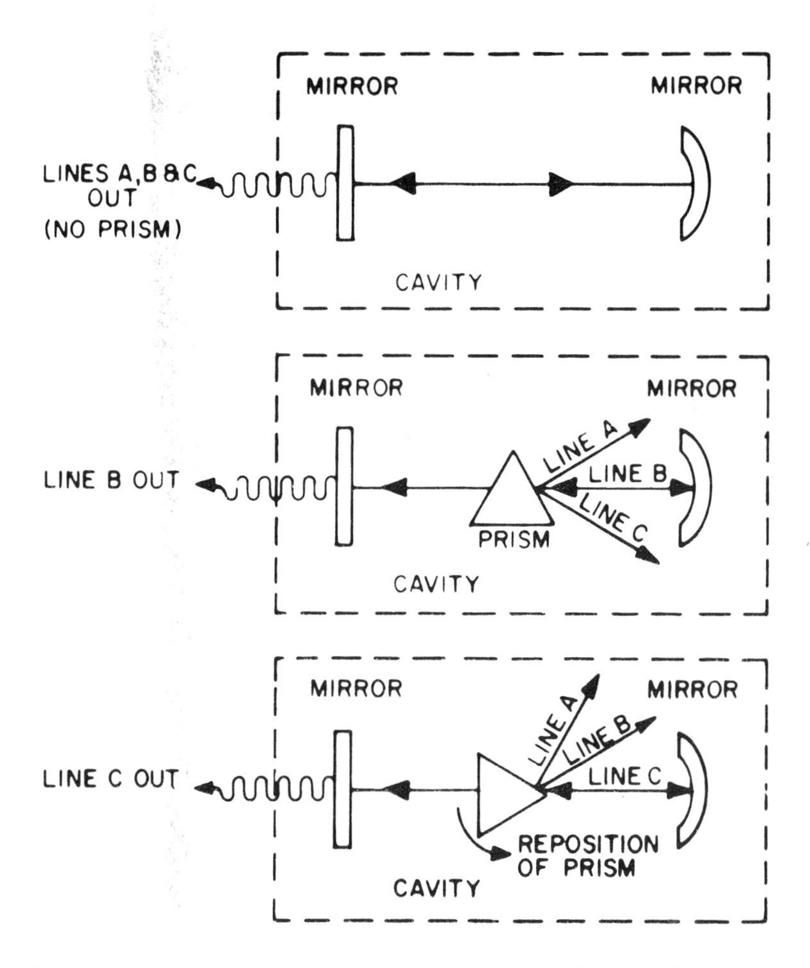

Some lasers provide only one wavelength, others provide many. These are known as multiline lasers. Such lasers can be made to provide each line separately by the use of a prism. (*Courtesy RCA*)

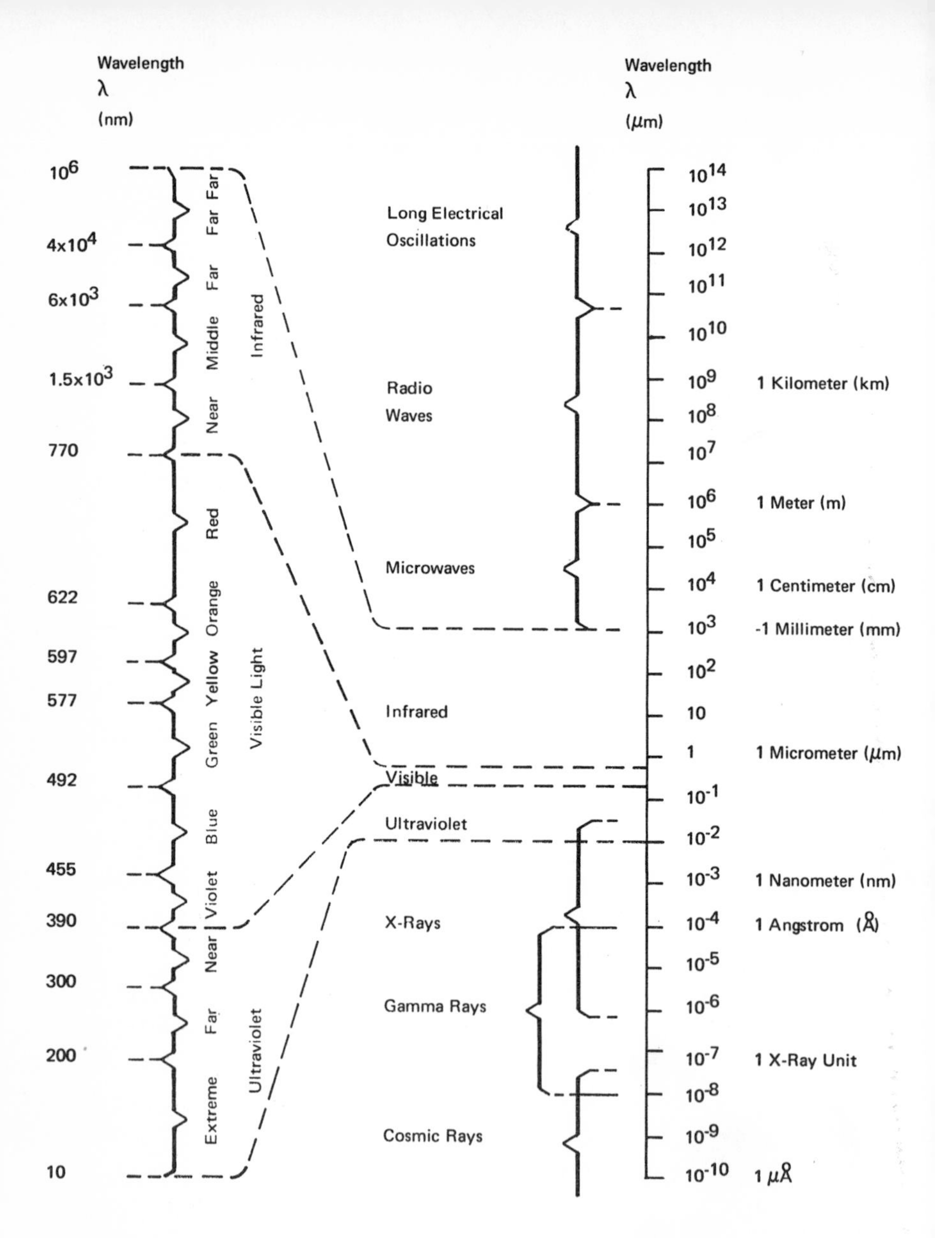

The electromagnetic spectrum. Notice the small area occupied by the visible light rays. *(Courtesy RCA)*

Lasers which are carefully adjusted to operate with a particularly stable pattern in the longitudinal and transverse movement are said to be mode-locked.

Frequency signifies the number of electromagnetic waves (crest-to-crest cycles) which oscillates past a given point per second.

The velocity at which a wave travels can be determined from the formula: Frequency x Wavelength=Velocity. In the case of electromagnetic waves in free space this speed is fixed, since the speed of light is a constant (c) which is equal to 186,000 miles per second.

All electromagnetic waves in combination make up the electromagnetic spectrum. Included are waves of light, radio waves, microwaves, infrared and ultraviolet rays, X rays, gamma rays, and cosmic rays. Each is differentiated by wavelength and frequency.

The spectrum ranges in frequency from about 10 cycles at the lower end for the longest radio waves to 10^{24} cycles per second at the top for cosmic rays. The longest radio waves stretch for thousands of miles. These are expressed in kilometers. Shorter radio waves are expressed in meters, being equal to several yards in length. Microwaves are measured in centimeters and millimeters. Higher up the frequency scale the waves shrink to microns (millionths of a meter), millimicrons, and angstroms (1/10 millimicron).

The wavelength band of lasers is approximately in the frequency range of 3,324 Å to 106,000 Å. The only portion visible to us is between 4,000 Å and 7,000 Å. The

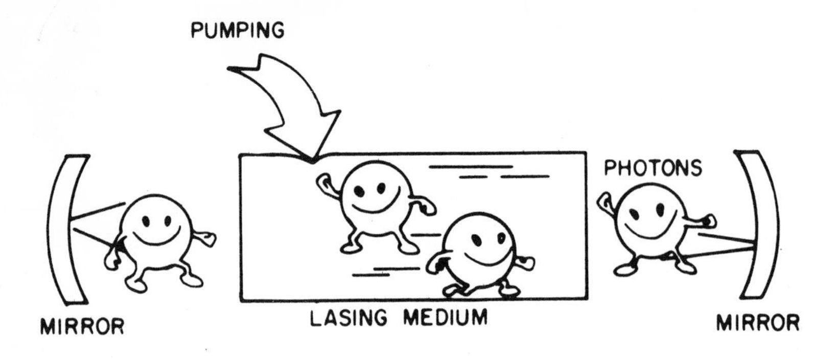

The basic components of a laser are the lasing medium (crystal, glass, liquid, gas, or semiconductor), the energy pumping source (electric current or light), and a pair of reflectors (mirrors or the polished ends of the lasing medium). The photon is the oscillating device of the laser. It is the energy given off by the atoms in their excited state. The photons establish the frequency (color) of the laser.

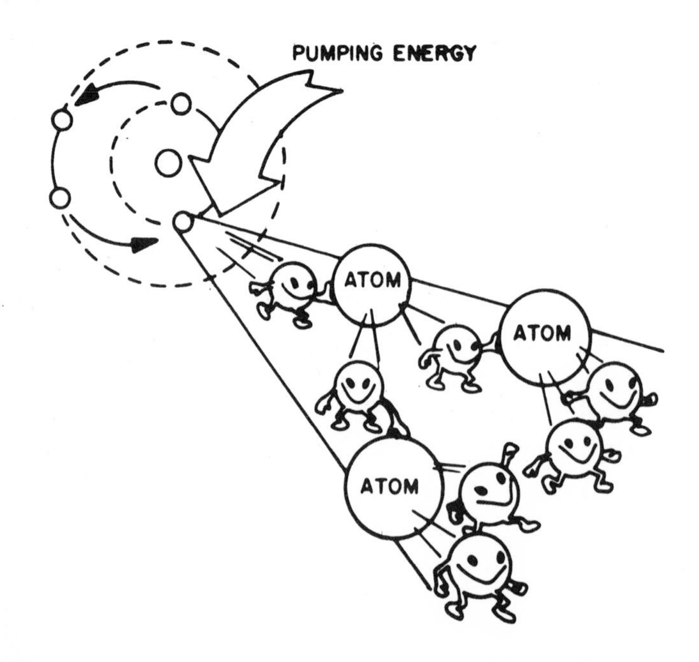

laser's electromagnetic waves oscillate between 4.23 x 10^{13} and 7.23 x 10^{14} cycles per second, commonly designated gigacycles.

The basic composition of all matter is its atomic structure, consisting of atoms or clusters of atoms called molecules. Each atom, in turn, is filled with plus- and minus-charged particles known as protons and electrons. The protons compose the heavy nucleus of the atom, and whirling around the nucleus are the negatively charged electrons. Each proton is balanced by an electron under normal atomic conditions.

The electrons orbit the nucleus at various levels of energy; the lowest is the rest level and the highest is the excited level. To move from the rest to the excited level, an electron needs a kick, something to energize it. Heat is one way to provide this boost; light and current are others and the ones with which we are concerned. When an atom is energized to the point where it releases the excited electrons from orbit, the atom as a whole becomes a charged particle known as an ion, and the electrons become "free" (unmatched with a proton), upsetting the neutrality of the atom.

Since atoms, or at least their electrons, are basically lazy, they prefer the rest state known as the ground level. In seeking this level, they must give back the energy which boosted them in the first place. They do this in the form of light or photons. The energy level arrangement in atoms differs, either in the distance between levels or in the number of levels. The result is that every photon takes on a specific frequency (color) based on the energy level arrangement of the atom. The colors are distinctive and enable scientists to "fingerprint" and identify unknown laboratory specimens.

To excite an atom so its electrons will jump to an upper level, the energy must be absorbed by the electron and not emitted. This energy must be equal to the energy difference between the lower and upper energy levels in the atom.

In all matter (group of atoms) there is always some activity between energy levels: up because of collisions between atoms and photons, down because of the electron's natural desire to seek the rest state. However, there are usually not enough electrons available at the upper level at any instant to drop to the lower level—hence, no light.

To make light, the majority of electrons must be on the upper level at all times so that when they drop to the ground level they give off an abundance of photons for absorption by the low-level electrons. Because this emission is random or "spontaneous," the electromagnetic waves of the photons are out of synchronization with each other–they are incoherent, such as we see in ordinary light.

To obtain the coherent light waves produced by the laser, the photons must be in step so that each wave travels precisely in a narrow parallel path behind the one ahead, crest for crest, trough for trough. This is the essence of laser operation and is the result of stimulated emission.

Stimulated emission means that the electrons on the upper level of the excited atoms are forced to give off their photons and are not permitted to release them at random. This is done by causing these atoms to be struck by a loose photon in that fraction of a second before the electrons are ready to plunge to the ground level on their own. Result: Two photons of the same frequency emerge from each stimulated atom—the one that struck the atom and the one produced by the fall of the atom.

When light is forced out of stimulated atoms by colliding photons, both photons are coherent—that is, the colliding photon and the one generated by the fall are in phase. Stimulated emissions make up about 10 percent of all light output, and it is this small portion of electromagnetic energy that composes 100 percent of the laser's light.

Most lasers are made of "guest" and "host" atoms. The guest atoms are chemically bonded to the host atoms by a process known as doping. For example, Dr. Maiman used a synthetic ruby as the host material and doped to it a mixture of chromium as the guest material. Doping must be done in a carefully controlled environment so that exactly the right number of guest atoms is connected to the host. The principle function of the host is to hold the guest atoms together in an orderly manner and to keep them from concentrating in one place to avoid lowering the energy level.

To keep the atoms working, "population inversion" must continually occur in the laser material. This is the elevation of the bulk of the atoms to the energy level—in effect, the turning upside down of the atomic concentration by means of the "pumping" action of the outside light source or current.

Also essential to laser action is a middle or "metastable" energy level. Here electrons on their way down from the energy to the ground level linger to stabilize momentarily. Although photons are not released as a result of this intermediate drop, energy associated with it is imparted to the laser material, causing the atoms to vibrate. When the electrons are stimulated to drop to the ground level a large concentration of photon energy is released.

Lasers are available for operation in two modes: (1) pulsed and (2) continuous wave (CW). The pulsed laser generally has less coherence because of the intermittent optical pumping but provides a much higher intensity light, hence more power. Most ruby and glass lasers operate in the pulsed mode, giving the highest instantaneous power outputs. Gas lasers can operate in either the CW or pulsed mode, although the best CW performance comes from gas lasers.

To coax more power from the laser than would normally be produced, a method known as Q-switching is used. This works, in effect, like an energy dam. A fast-acting shutter is placed between the laser rod and the emitting end mirror. Closing the shutter prevents the laser

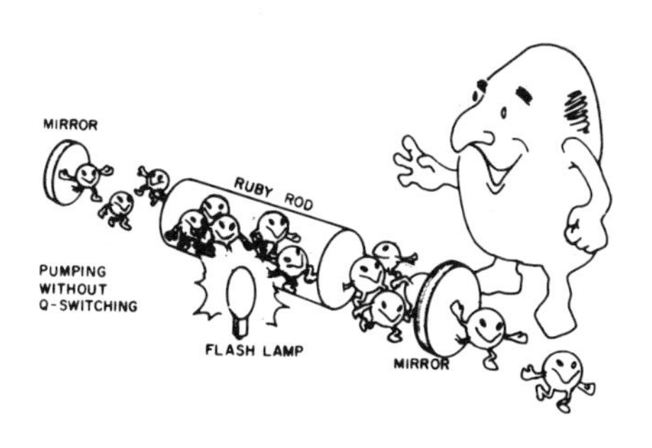

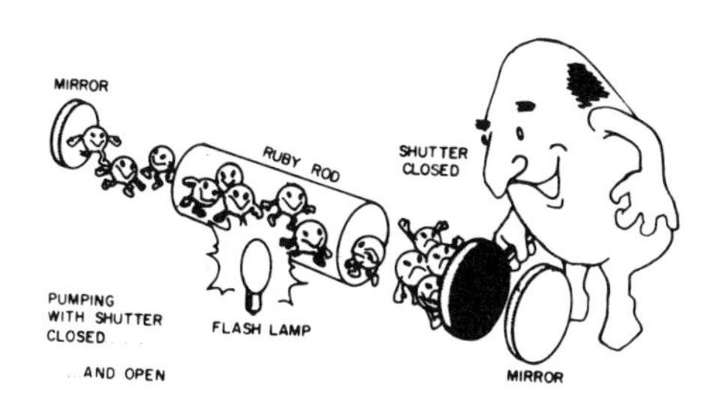

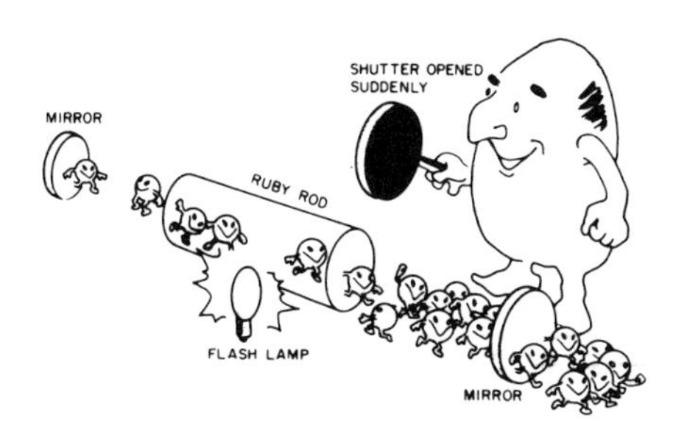

Normally when the laser is pumped light energy in the form of photons is emitted as soon as population inversion is high enough for lasing. In a Q-switched laser the light energy is blocked from breaking into oscillation immediately by means of an optical shutter in order to increase the pulse. (*Courtesy RCA*)

from breaking into oscillation at the moment of population inversion threshold. Although the laser is excited to the point where it would emit coherent light, the closed shutter keeps this from happening by cutting off photon cross fire in the laser cavity while pumping continues. As a result of this inhibition and the storing capacity of the laser, the metastable level becomes vastly overpopulated. When overpopulation is at maximum, the shutter is opened, radiation builds up rapidly, and the abundance of stored energy is released in a giant pulse.

This method is also termed Q-switching, *Q* standing for the quality of the laser rod as a resonant cavity which is "switched" or interfered with while the shutter is closed. Subsequently the need for a shutter was eliminated in some lasers by the devising of a means of rotating the emitting end mirror, so that by turning it so it was not parallel to the fixed mirror at the opposite end of the resonant cavity stimulated emission was blocked.

Q-switched lasers have achieved beam energies of thousands of joules and peak pulse powers of many hundred millions of watts. Unfortunately, this method of generating coherent power is inefficient and impractical because of the vast amount of optical pumping energy needed to achieve it.

3
Building a Laser

The laser, like the electronic oscillator, is an amplifier with feedback—in this case a light amplifier. Similar to the oscillator, the laser requires an oscillating medium, an external energy source, and a resonant circuit with sufficient feedback to sustain oscillation.

The oscillator of the laser is the photon, which is produced by the energy level changes of the electrons within the atom. The external energy source of exciting the atoms may be either optical, in the form of a flash lamp, or electrical current.

The laser's resonant circuit consists of the lasing medium (solid, liquid, or gas) and a pair of properly adjusted reflecting surfaces at either end of the lasing material. The light reflectors may be a set of mirrors or polished ends of the laser material, such as with the ruby.

Since the ruby is probably the most commonly used laser material and that from which Dr. Maiman built the

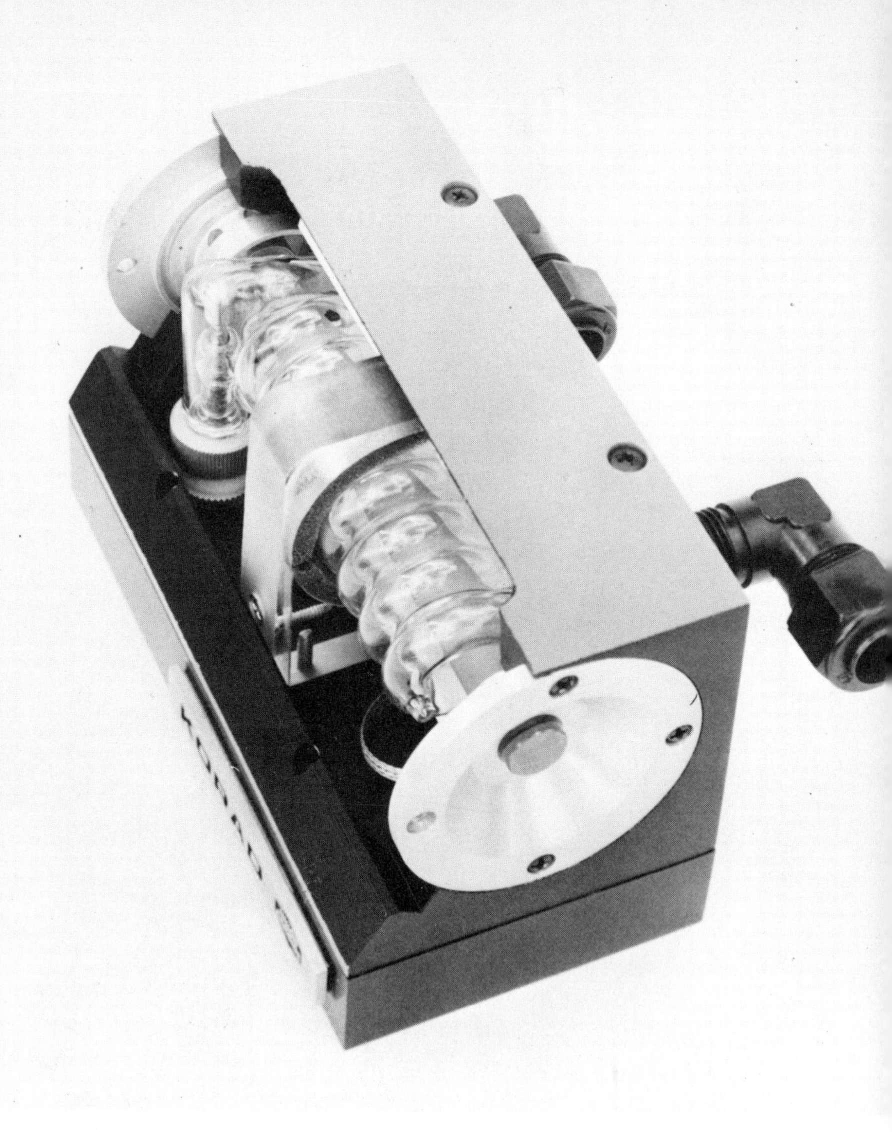

Cutaway of a Korad laser cavity showing the spiral xenon flash lamp, the ruby laser rod, and the cover. (*Courtesy Korad Department/Union Carbide*)

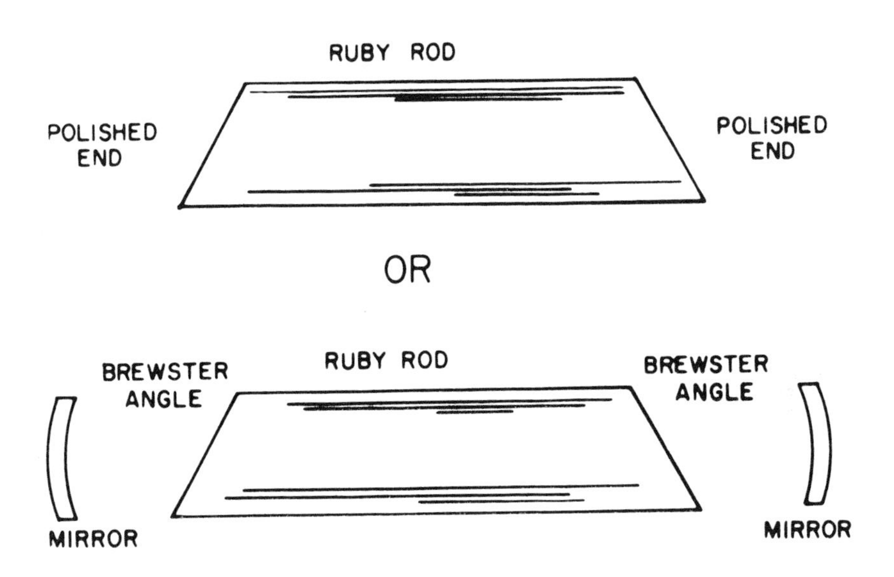

Solid lasers, such as the ruby, may use either their polished ends or separate mirrors as reflectors. (*Courtesy RCA*)

world's first laser, it seems appropriate to describe how the ruby laser is made.

A piece of ruby crystal has to be "grown" under carefully controlled conditions and cut to exact size. The growth starts by dipping an aluminum oxide crystal into a mixture of chromium oxide and aluminum oxide atoms in prescribed amounts. Bubbles, lumps, and any other imperfections in the mixture have to be removed, since these will affect the material's lasing capability. You will recall that Dr. Maiman's initial synthetic ruby failed to operate because of these imperfections in it.

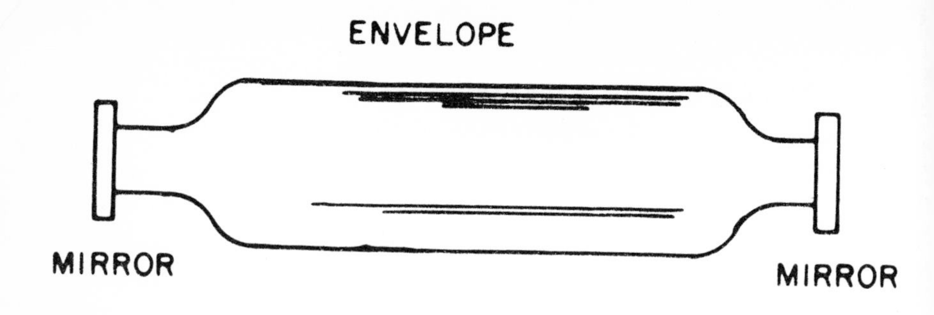

OR

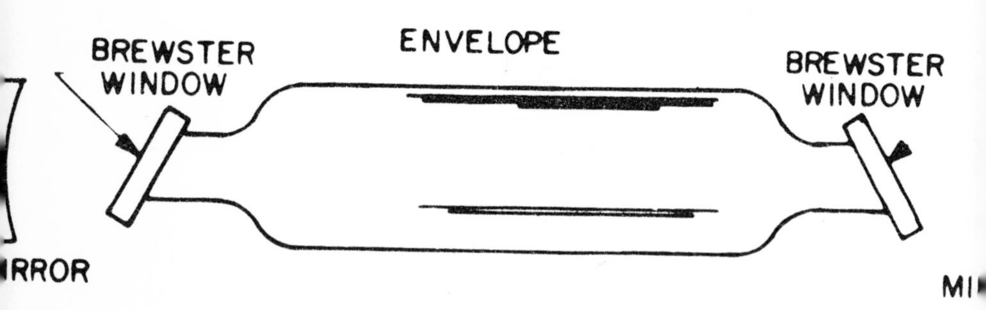

Gas and liquid lasing media must be contained in an envelope. The cavity mirrors may act as seals to the envelope, or they may be separate from the envelope with Brewster angle windows serving as the seal. (*Courtesy RCA*)

The raw cluster of ruby crystals must be removed from the growing solution at precisely the right moment. This raw mass is in the shape of a pear and has been named a boule, which means bowl- or pear-shaped. From the boule the laser rods are cut into the finished sizes desired.

Before the rod is inserted into the laser cavity, the ends must be extremely smooth, level, and parallel, then polished to a high degree of reflectance or coated with a mirrorlike substance. One mirror is totally reflecting and the other is partially so to emit the energy.

Most ruby lasers are optically pumped by a flash lamp, tungsten filament, or another laser. Dr. Maiman used the flash lamp, which he direct-coupled to the ruby rod by wrapping it around the rod so the light shone directly into the material. Another common method of coupling is known as image coupling, in which the light is reflected from another surface onto the ruby rod.

The center of the laser is sometimes called the optical cavity because it houses the electrooptical elements, such as the flash lamp and the lasing material.

The laser that Dr. Maiman built was encased in an aluminum cylinder. The synthetic ruby was shaped as a thin rod about one inch in diameter. The electronic flash lamp was direct-coupled to the rod. Each end of the rod was silver-coated, with one end partially transparent for the emission of the coherent light. In addition, the ends of the rod were ground to "optical flatness" so they would reflect only photons traveling in a direction parallel to the long axis of the rod.

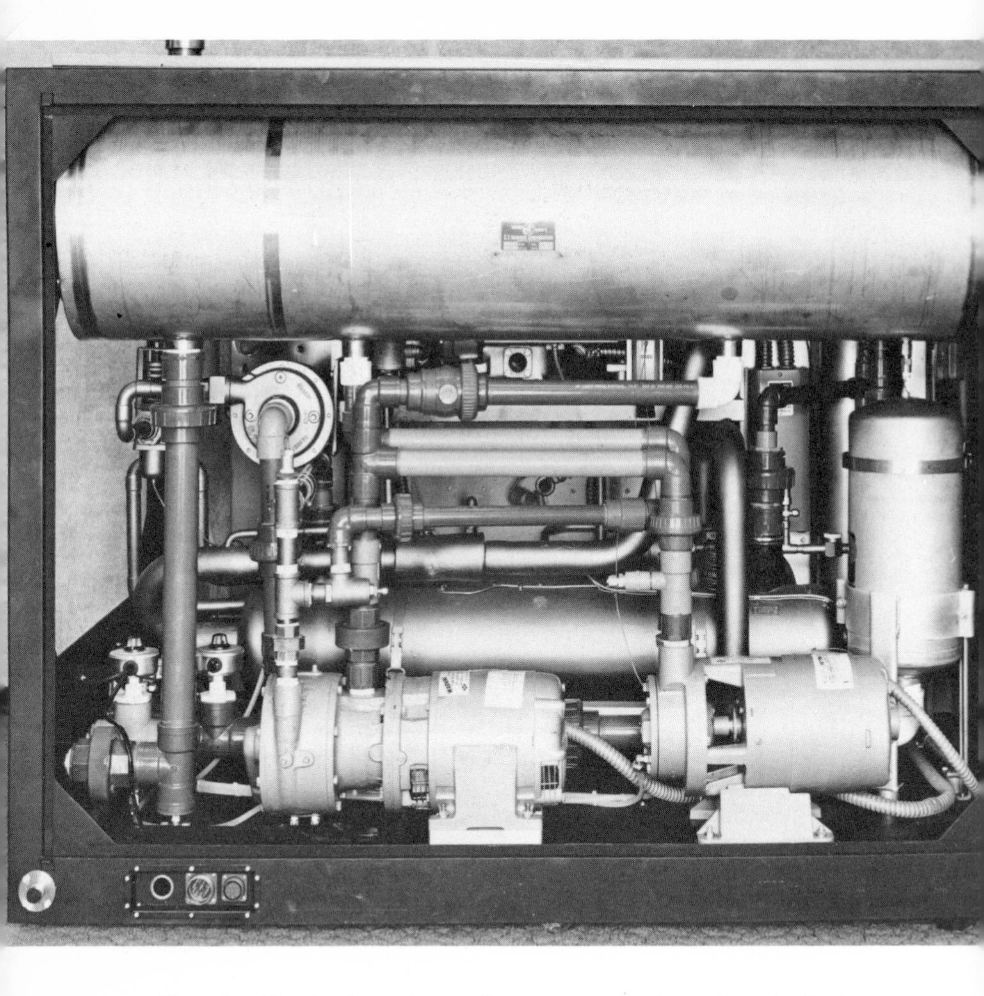

The plumbing in the water-cooling system of a large Korad ruby laser. (*Courtesy Korad Department / Union Carbide*)

When the flash lamp was lit, photons entered the ruby, striking the chromium atoms (the guest atoms) and exciting them to jump up to the energy level. The difference in atomic levels between the ground level and the energy level was equivalent to a photon with a wavelength of 5,600 Å. This wavelength corresponded to the green region of the visible spectrum and matched the green light which the flash lamp provided for the pumping energy.

From the energy level, the electrons dropped to the metastable level, and from here a few electrons spontaneously dropped to the ground level, releasing photons with a wavelength of 6,943 Å, deep red in color. At this point the ruby began to glow or "fluoresce," the same as ordinary light. As the photons that had been released simultaneously began to move back and forth between the mirrors they collided with the bulk of the electrons still in the metastable state. These electrons had been stimulated to emit together, and the few simultaneously released photons had now been transformed into many coherent light particles (quanta). As the process repeated itself and the energy continued to build up, suddenly the laser threshold was crossed and a brilliant flash of red light at 6,946 Å was emitted. Once the chromium atoms were back at ground level again and another flash of light excited them, another pulse was formed and emitted, and the laser continued to send out its awesomely brilliant light.

4

Laser Comparisons

In efforts to increase the output and efficiency of the laser, scientists have experimented with a wide variety of laser materials, explored different power sources, and assembled components into a number of configurations.

From one laser source, the ruby, they have produced countless sources within an ever-expanding array of basic laser types. These include crystal, glass, liquid, gas, and semiconductor materials.

Because of its optical purity, glass is an ideal laser material. It has a "relaxed nature" in terms of its electrical and mechanical characteristics. The precise matching of host and guest atoms is not critical in glass lasers. It wastes less energy than most lasers in the form of heat and lases quicker than most other types. Glass can be conveniently made into many shapes and sizes, and because of the flexibility between guest and host atoms, any number of guest materials may be used with glass.

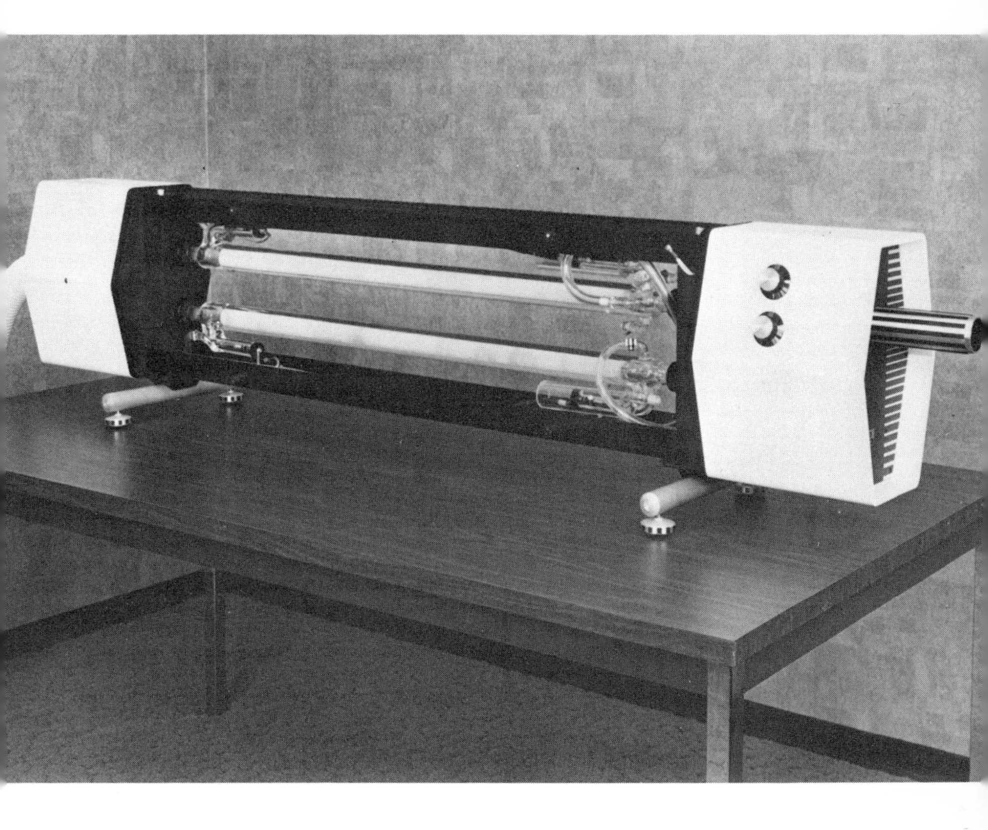

Cutaway of a continuous wave carbon dioxide gas laser built by Korad. (*Courtesy Korad Department/Union Carbide*)

By a method known as sensitation atoms of two crystals may be doped to glass together, lowering the energy threshold and thus further increasing the glass laser's efficiency. Operating in the pulsed mode, glass lasers can store energy 10,000 times longer than injection (semiconductor) lasers and 10,000,000 times longer than gas lasers. Glass, along with ruby lasers, produces an intense, blinding power output, which makes it practical for microspot welding and drilling. Emission is in the infrared region and its efficiency is one percent.

Liquid lasers are made from organic and inorganic materials of a wide variety. They are usually contained within a glass envelope. They offer the unique feature of cooling by circulating the lasing medium. Liquid lasers can be made longer than solid lasers, thereby raising the total energy output and coherency.

In the pulsed mode, organic liquid lasers provide one megawatt of power. Emission is in the infrared and visible regions of the spectrum.

Inorganic liquid lasers emit at 1.06 microns (infrared) and in the pulsed mode provide up to 10 megawatts of power.

Efficiencies of the glass lasers range from 0.2 percent to 40 percent, if the laser is pumped.

One of the earliest to be conceived, the gas laser is in most common use in R&D applications today. Included in this group are such gases as argon, krypton, neon, helium-neon, carbon dioxide, and vapor types such as helium cadmium. Although the instantaneous peak

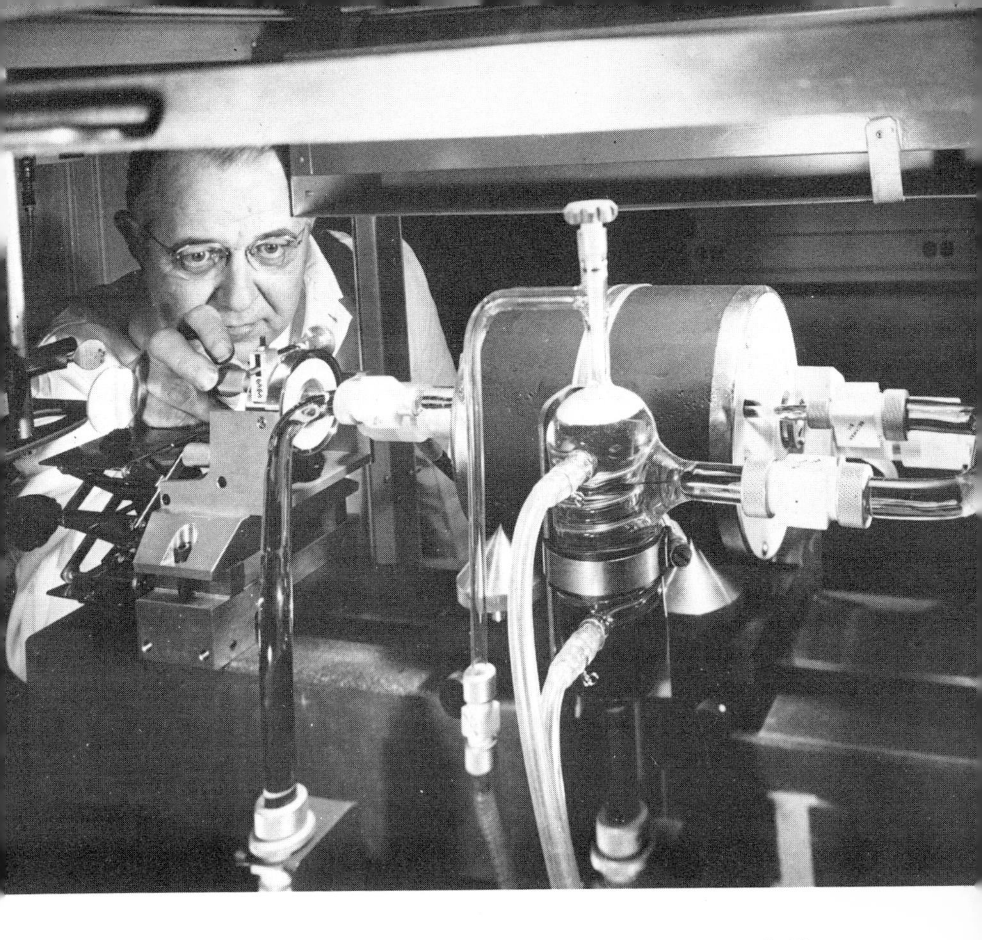

A circulating liquid laser developed by the General Telephone & Electronics Laboratories, Inc. Circulating liquid increases the speed at which the light pulses are emitted by the laser and makes possible more precise aiming of the pulse for future communication and display systems. (*Courtesy General Telephone & Electronics Corporation*)

powers of gas lasers are many orders of magnitude lower than glass, liquid, or crystal lasers, they do offer narrower band widths, less spreading of the beam over a distance, and high spectral purity and coherence. Gas lasers produce ultrapure colors and far exceed nongas devices in their coherent beams. Gas lasers, like liquid lasers, are able to assume almost any shape and size, being limited only by the flexibility of the glass containers.

The helium-neon laser, because of its simple construction and lower cost, is probably the most extensively used laser today. It works on the same basic principle as the ruby laser we described previously, although the process varies slightly.

A tube is filled with helium-neon gas in a certain prescribed ratio (10 to 1 or 5 to 1, with the largest volume of gas being helium). Each end of the tube has the customary reflectors for bouncing the photons back and forth for emission. Fifty watts of power, maximum, is needed to get the helium-neon mixture to lase and keep it operating. Because of this low energy requirement, little heat is generated and the gas laser can be worked at room temperature. The pumping power, instead of coming from a flashtube as in the case of the ruby laser, comes from a direct current (dc) generator. Several metal bands are placed around the tube and then connected by wires to the dc generator, which when turned on sends a steady stream of electrons through the gases and starts the laser action.

Another variation exists in the gas laser process known

This Hughes Aircraft Company ruby laser is vaporizing a .020-inch-diameter crater at approximately 4,800° F into the surface of stainless steel. (*Courtesy Hughes Aircraft Company*)

as dissociative excitation transfer. This involves the mixing of oxygen with certain gases like neon and argon. The operation depends on the fact that gaseous oxygen is normally molecular, composed of two oxygen atoms locked together chemically. When a dc discharge passes through the laser tube, the energy imparted to the neon or argon atoms lifts them to the energy state. The atoms then collide with the oxygen molecules on the ground level and the impact splits (dissociates) the oxygen molecules into separate atoms. The energy from the collision is transferred to one of the atoms of oxygen which then rises to the energy state, while the other oxygen atom stays on the ground level. From here on the stimulated emission process is the same as for other gas lasers.

Still a third modification of the basic laser process should be noted with the gas laser. This time only a single gas is used, any one of the so-called noble gases: helium, argon, krypton, neon, and xenon. These gases lase through a direct collision between the high-energy electrons carrying the dc discharge and the gas atoms. From the energy level, the gas atoms fall en masse, as a result of the stimulated emission, and produce continuous, coherent beams on many hundreds of different wavelengths spanning the entire visible spectrum, the ultraviolet, and infrared.

With the noble gases so much collision energy can be given to the atoms that several electrons orbiting each atom are knocked free of their orbit simultaneously,

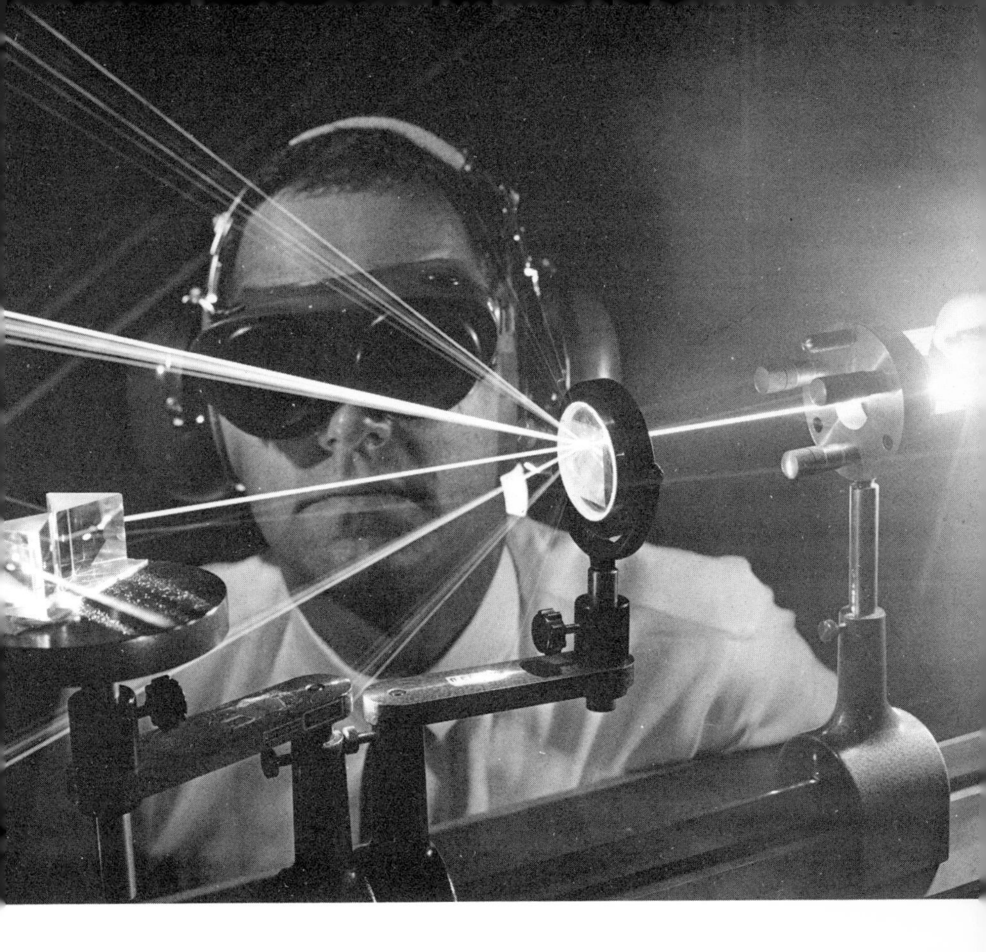

The argon gas laser, invented by Hughes Aircraft Company, can be used for space communication because of its extreme brightness and narrow beam. It also has use as a laser telephone, carrying millions of voice and TV signals, and as an optical reader for computers. (*Courtesy Hughes Aircraft Company*)

leaving positively charged "ions." The electrons remaining in the ion assume different levels, and this enables the attainment of the many different wavelengths. Some of these wavelengths are in the blue-green region of the spectrum and, unlike other visible colors, are not absorbed by water. This discovery has led to the development of laser underwater communications and tracking systems.

Argon gas lasers are finding increasing acceptance in such precise tasks as Raman spectroscopy, holography, alignment, and information handling because of the blue-green characteristic of the lasers.

The krypton lasers, although of lower power output, can provide a broad spectral range.

Carbon dioxide lasers offer the highest peak power and efficiency of all gas lasers (about 15 percent versus 0.1 percent) but are limited to operation in the far infrared region—10.6 microns.

With the addition of vapors, such as optically pumped cesium, to the list of gas lasers these devices are able to put out photon energy in the submillimeter region of the spectrum, which offers a whole unexplored area of potential applications.

Semiconductor (injection) lasers offer the highest efficiency of all lasers, as high as 70 percent at liquid helium temperatures. However, radiation from these lasers spreads out more than that of other solid type and gas lasers, so they cannot be used in applications where beam divergence must be at a minimum.

Key to the operation of the semiconductor laser is the amount of electrical current injected. It has to be enormous, or below a certain threshold level the photon emission is scattered, lacking coherence and intensity.

"Semiconductor" describes the kind of material from which the laser is made, a fair conductor of electricity. The best known material is gallium arsenic (GaAs), which in typically pulsed operation gives up to 100 watts of energy at 0.8540 microns in the infrared region. In the CW mode it gives 1 to 10 watts at 0.9040 microns (IR) at 77° Kelvin (liquid nitrogen cooled).

Laser action takes place in the 1/10,000-inch-wide junction or buffer zone between the positively charged electron "holes" and the electrons in the semiconductor material when the device is connected to a battery. The battery injects a powerful current into the semiconductor and this, in turn, excites the GaAs atoms so that electrons and holes cross the junction. The electrons drop into the holes and lose their energy, which is released as photons, and the usual laser pattern follows. The intense narrow beam is no wider than the minute junction.

During operation, the semiconductor laser is immersed in a supercold helium or nitrogen liquid, although recent improvements have made it possible to operate some of these devices continuously at room temperatures without burning up.

The efficiency potential of semiconductor lasers is an amazing 100 percent. Its beams can be easily modulated—that is, carry information. By varying the current which drives these lasers, it is possible to lessen or increase the intensity of the emerging beam. If the current variations correspond to the data to be transmitted, so too will the resultant changes in beam intensity in order to convey voice, music, and pictures.

These are the materials scientists have successfully applied to the laser. But theoretically the laser can be made of any substance that gives off light. So it is possible that someday soon this may consist of an ordinary light bulb which we buy at the supermarket.

5
Laser Applications

Today, most scientific resources are being directed toward laser applications. Since the announcement of the laser in 1960, many people have either forgotten about it or lost interest in its "magic." Among technical men and women criticism has grown, since it was felt that the laser has not lived up to its expectations as "the light fantastic" that would benefit every area of human endeavor.

Before the laser can become the universal technological tool of man, many difficult problems of efficiency and control have to be solved. These problems are being worked on by men of vision and patience whose confidence in the ultimate usefulness of the laser is evidenced in the applications already available to us.

Basically, the laser is used to perform tasks which conventional tools cannot do as effectively. That seems to be the criterion of selection today in the predominant areas of laser operation: scientific experiments, com-

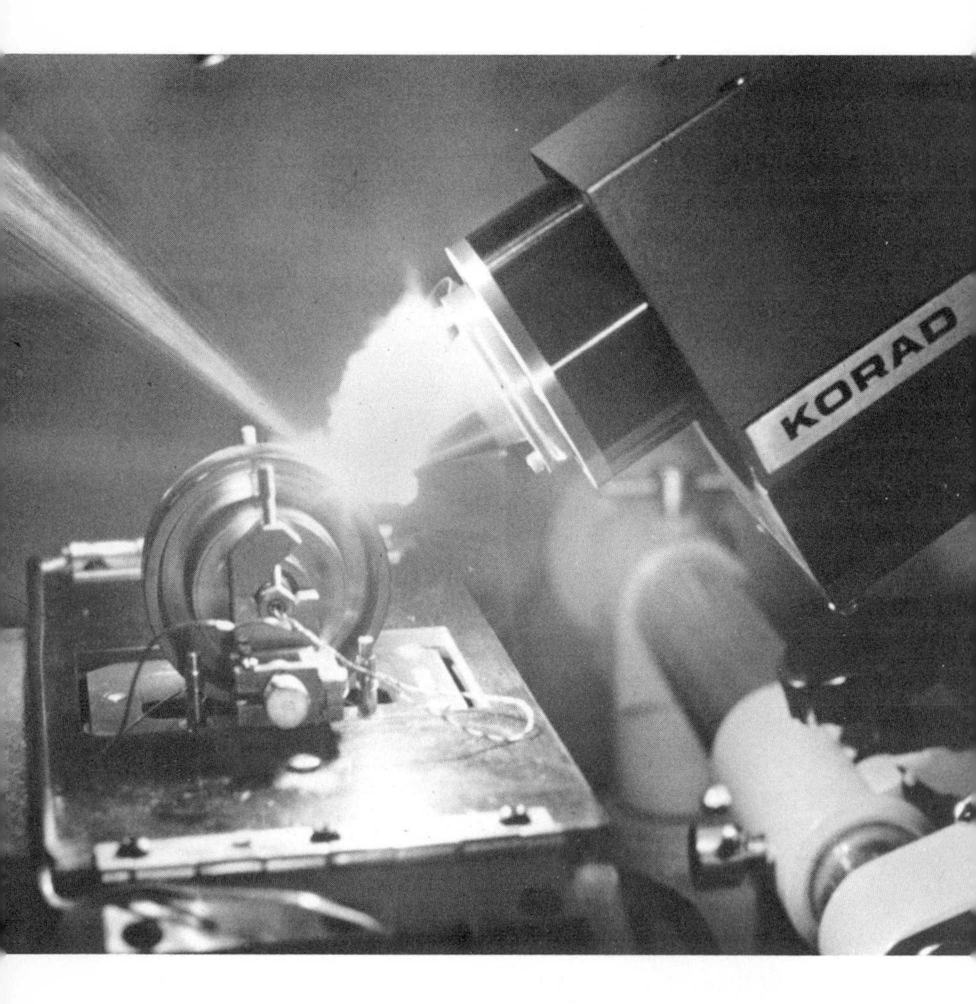

Trimming a gyroscope for perfect balance with a Korad ruby laser. (*Courtesy Korad Department/Union Carbide*)

munication, navigation, medicine, industry, and navigation.

The laser began as a laboratory instrument, and it is here that most of the scientific breakthroughs in coherent light technology have taken place. The laser has made significant improvements in interferometry and measurement standards, spectroscopy, and the study of harmonics.

The coherency of the laser's beam is its key to unlocking the previously inaccessible area of extremely precise measurement. Because of its exceptional purity, the gas laser works best in this application. So accurate is this laser that it can measure the distance between "invisible" atoms, as well as other imperceptible parameters. It does this by an extension of the capability of the Fabry-Perot interferometer. The laser uses a series of mirrors that reflects the light beams from the atoms as they bounce back and "interfere" with each other. It then counts the surface lines between the light beams and determines the distance in millionths of an inch—about 1/250 of a human hair's diameter.

The replacement of incoherent light sources in the interferometer with a laser greatly extends the usefulness of this instrument. Previously coherency could be maintained for only a few feet with the interferometer; now this range has been extended to several hundred feet by the laser. This extended capability enables the interferometer to measure the power output of large antennae used in space tracking and communication, on one hand, and the minute thickness of mirror and lens coatings, on the other.

The laser may eventually become the instrument used to set the world's standard of length. It can detect previously "unmeasurable" changes in the length of an object and has accurately detected eight parts in several billion. This is like finding a pebble in a ten-thousand-mile-square desert.

The laser is ultrasensitive to earthquakes. Two gas lasers are being used at the University of California to verify the severity of earthquakes and nuclear explosions. The two lasers are lined up end to end, and each has a mirror attached to a weight on the end of a spring. When a disturbance occurs, the spring and mirror arrangement moves, shifting the frequencies of the two lasers—one up and one down. The new beams are then relayed to a special crystal which mixes them and emits a new frequency equal to the difference of the original two laser frequencies. This difference is a direct measure of the amount of the force in the earth's movement caused by the earthquake or nuclear blast.

Lasers are aiding in the identification of substances by spectroscopy. A spectrometer vaporizes the substance and enables it to be identified by the color of the wavelength emitted on a screen. Use of laser pulses to heat the substances has simplified the sample preparation task and has made it possible to analyze the most minute samples.

Previously impractical with existing instruments,

A portable ruby laser range finder developed by Korad. (*Courtesy Korad Department/Union Carbide*)

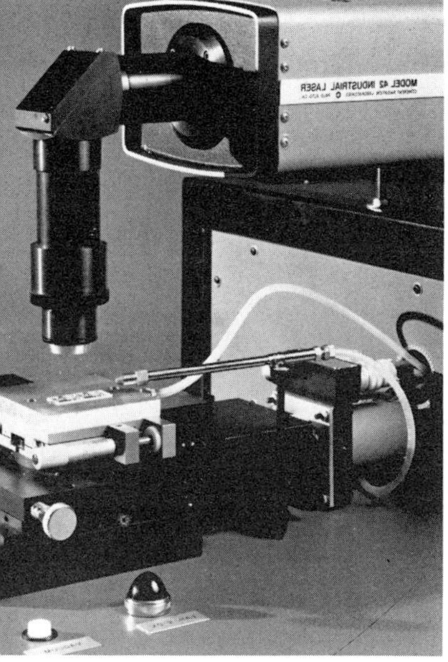

An example of an industrial gas laser in use today for cutting, drilling, scribing, and soldering a variety of materials. The CO_2 system was developed by Coherent Radiation. It has applications in the microelectronic, glass, ceramic, plastic, and textile industries. (*Courtesy Coherent Radiation*)

Looking like a three-legged robot, this laser surveyor, built by Laserplane Corporation, projects a thin disk of light 2,000 feet in diameter with extreme accuracy. The height of the disk is electronically read by a detector on the surveyor's rod. (*Courtesy Laserplane Corporation*)

harmonic sound waves can now be studied effectively in liquids and solids using a laser. Again, this is achieved through the observation of frequency shifts in the laser as the beam enters and leaves the material, resulting in a Raman shift effect, a harmony in the rise and fall of the frequency.

Lasers are used in nonlinear optical studies of materials. Although scientists had been aware of the nonlinearity possibility of certain substances, they had not been able to test for it. By focusing laser light on a substance, if nonlinear, the emerging beam will be twice the frequency or some other harmonic of the original fundamental frequency. These harmonic generators, as the crystals are designated, can boost the number of laser wavelengths available.

The laser has been used to verify the fact that the speed of light is a constant 186,000 miles per second. The ammonia gas maser was used as an atomic clock with an unbelievable accuracy of an estimated loss of one second in ten thousand years. These experiments were first steps toward establishing the laser as an independent time standard against which all time can be computed.

The laser beam has been used in the laboratory to produce plasma gas and to measure the electron density distribution of this extremely hot gas. Previously, studying this gas was a problem, since any mechanical device placed in it would disintegrate. The laser stabs a coherent beam into the plasma, and again, changes produced in the frequency by the molten mass enable scientists to measure the electron density of the plasma.

A laser beam can make chemical reactions start. It can be focused through a high-powered microscope to study microbes, viruses, and cells. The beam can be focused to pinpoint sharpness, pick off a tiny piece of the cell, and help clear up the mysteries of growth.

Today the laser is opening a whole new realm of communication because of the vast information-handling capacity of the coherent beams. It has been estimated that a single laser beam can carry all the data transmitted by every radio and telephone in the world today—simultaneously. Unfortunately, there are certain serious obstacles that must be overcome before the laser will eliminate the telephone pole and other conventional communication media. Laser light, like any light, is absorbed by the moisture in the atmosphere. Thus in bad weather transmission is unreliable, perhaps impossible in some situations. Another problem, which is gradually being solved, is the ability to modulate and demodulate the information—getting it in and out of the laser beam.

Since space is free of atmosphere, the initial use of laser communication is in satellite and spacecraft communication. The laser is also expected to overcome the hazards of communication blackout during spacecraft reentry by being able to pierce the ionization barrier produced by the heated nose cone.

Television by laser is still experimental, although it was successfully and dramatically demonstrated at the

World's Fair in Japan, Expo '70, on a huge nine-foot by twelve-foot screen. Laser TV uses a pencil-thin transmitter and receiver which are interchangeable and which project the image on a screen, rather than on a tube as in conventional sets. The normal light of the environment is fine for illuminating the subject, and the laser beams a beautiful, sharp color picture, with colors derived from different beam frequencies.

Tracking or ranging an object to locate its exact position is another job being done by lasers today. U.S. Army laser rangers fire high-energy beams over short distances at a target. The time it takes for the light beam to reach the target and return is fed into a computer built into the ranging system. Based on the total number of pulses of the round trip, the exact distance and location of the target are given. Present distance is limited to fifteen miles within the atmosphere because of interference from rain, fog, and smoke. In space, however, the laser ranger can accurately measure and locate targets at great distances. For example, a laser range finder measured the distance from the earth to the moon within six inches. Previously, the best measurement was one-fifth of a mile.

By the same principle, lasers have been used to map the bottoms and rims of the moon's craters.

Laser beams have replaced the flow of electricity in some experimental computer optical readers, enabling these devices to scan, identify, and reproduce information with the fabulous speed of light. For example, the laser has been able to locate in a millionth of a second a single word on a long roll of microfilm. Conventional means of searching, finding, and printing the same word would take seconds, and even though this seems fast, this retrieval process actually slows down the computer. In addition, fairly exotic all-laser computer systems, operating on altogether different principles than those of computers in use today, are in development.

With laser pulses as the light source for high-speed cameras, photographers are now able to shoot 500,000 pictures a second. The application is primarily in military aerial reconnaissance and mapping.

Also, through another revolutionary process known as holography, lasers are replacing lenses and making three-dimensional photographs. Holography has proven to be extremely useful in nondestructive testing, particle studies, and wind tunnel experiments. It has been used to analyze sandwich structures, composite materials, tires, and rubber-to-metal and metal-to-metal bonds. The structures are studied for subsurface defects. Holography can measure the size and shape of suspended particles in air or liquid. It permits optical examination of rapidly changing aerodynamic phenomena in wind tunnel experiments. Even shock waves from bullets have been made visible.

Holography is the technique of recording the image of an object using the entire content of the light reflected or transmitted by the object. The word "holography" is derived from the Greek *holos* and *grapho* meaning "to write whole." "Whole" refers to the entire content of

A holocamera looks into the rocket exhaust nozzle during a test at the Jet Propulsion Laboratory-Edwards Test Facility, California. The holocamera provided scientists with three-dimensional holograms of the engine's combustion for study in order to improve the engine's performance. (*Courtesy Jet Propulsion Laboratory*)

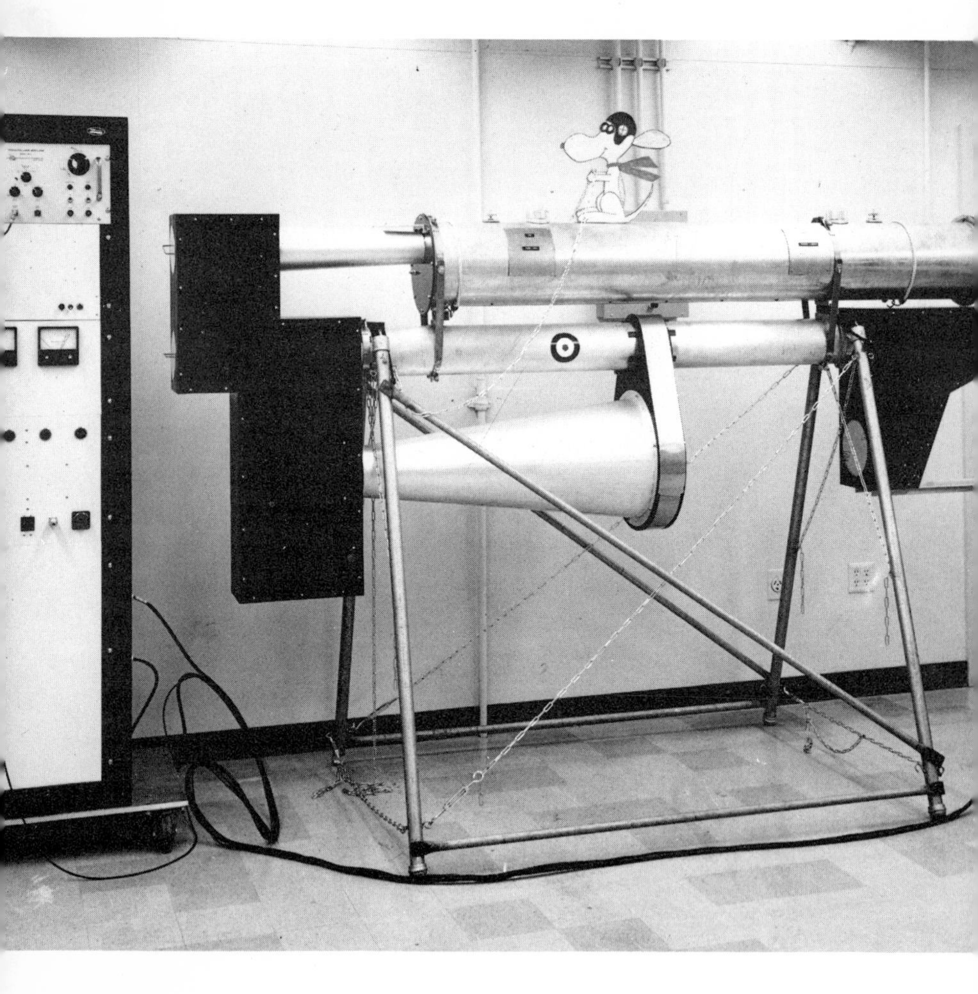

A transmission holocamera developed by TRW Systems to record the flame combustion in experimental liquid rocket engines so scientists can improve performance. (*Courtesy NASA/JPL*)

light: intensity, wavelength, and phase. Conventional photography is only concerned with part of the content of light: intensity and wavelength. Through the use of lenses and mirrors, it records the image by the intensity distribution of the light reflected by the object.

Holography makes use of the phase content of light to replace the optical elements. The mutual position of the parts that constitute the object is rendered by the phase of the light waves reflected or transmitted by the object.

When two waves of the same wavelength arrive together at a certain point interference occurs. It is constructive if there is an increase in intensity at this point. It is destructive if the waves cancel each other and leave darkness.

To produce a hologram the laser beam is separated into an object-illuminating beam and a reference beam. The two beams are obtained from the same laser by a beam splitter. The reference beam is reflected directly into the photo plate, and the laser light reflected by the object intersects and interferes with the reference beam. A photographic plate inserted into the interference pattern and properly exposed will record the pattern. After development the plate is called a hologram. When monochromatic light is later passed through the plate, its waves pulse with the many interference fringes on the plate and reproduce the original scene in three-dimension.

Lasers as navigators are already in operation in gyroscopes and radar systems, altimeters, and underwater

Photographs of reconstructed holograms of R. A. Briones of the Electro-Optics Group of TRW Systems. The holograms were recorded with a ruby laser on a photographic plate. The complex light wave patterns of the laser-illuminated subject were reformed in the photographs shown here when the monochromatic light of a helium-neon laser was passed through the photographic plate. (*Courtesy TRW Systems Group, TRW*)

sonors. In gyroscopes the laser replaces the mechanical spinning device for systems of unparalleled accuracy. Most commonly in use for this application is the gas laser, whose beam spins around a series of carefully placed mirrors. Built-in detectors measure the frequency shifts of the beam as it spins, so the laser can adjust instantly to the correct course no matter how complex the maneuver of the vehicle.

Laser "optical" radar systems are becoming increasingly important in space exploration. Laser radar can be focused into a beam at least one hundred times narrower than microwaves, and each beam of coherent light is much more accurate in locating objects and tens of thousands of times faster in telling the speed of moving objects. Laser radar systems use drastically smaller antennae—inches versus feet.

These optical radar systems have been used for fog and smog detection and for finding "invisible" atmospheric layers. They have been able to see areas of turbulence that could be dangerous to aircraft. Laser radar pulses have measured cloud heights and drift with great accuracy.

Because the narrow beam of optical radar can spot objects too small for conventional radar systems, it is finding increasing acceptance on small craft navigating inland waterways. It can pinpoint partly submerged objects in rivers and bays and has proven to be a good collision-avoidance device in crowded harbors.

One of the laser application areas which most

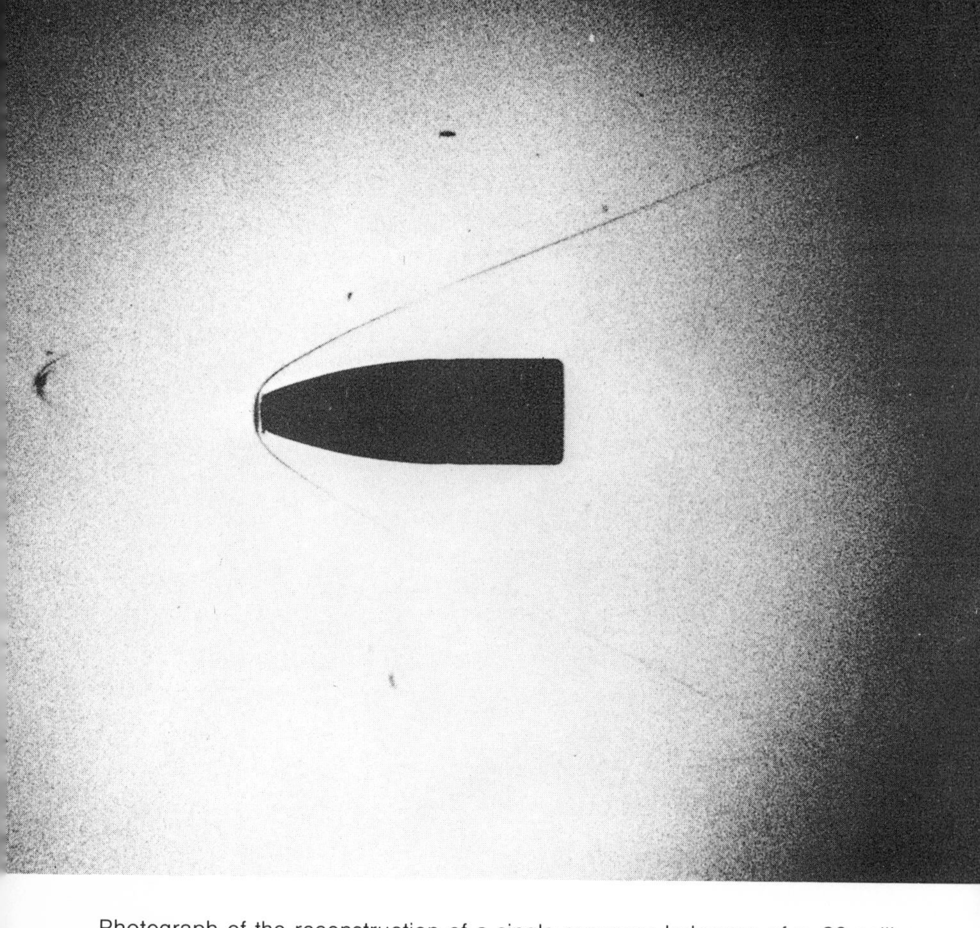

Photograph of the reconstruction of a single-exposure hologram of a .22 caliber bullet in flight at 3,500 feet per second. (*Courtesy TRW Systems Group, TRW*)

stimulates the imagination is that of medicine. We can envision laser knives so sharp that they do not cut the skin, coherent light of intense heat that fuses the cavities of our teeth and makes them more sparkling than when they were new. We can imagine invisible healing rays that remove all the world's ills in a flash. It is quite possible that these laser "miracles," forecast originally in the early 1960's, will become realities in the next decade. But for now the laser is still primarily an experimental medical instrument, useful for a few tasks in the hands of skilled surgeons.

For example, a microscopic beam of ruby laser light is used to "weld" a detached retina in a thousandth of a second. The operation can take place in the doctor's office and without any discomfort to the patient. As soon as it is over, the patient can go home. In the past, with conventional surgical methods, the operation could have taken hours, and the patient would have had to spend weeks in the hospital or at home recovering. The powerful beam of the laser literally welds the eye tissue together, making the connection stronger than it was originally.

The application of the laser to such diseases as cancer is only beginning. Surgeons, using lasers the size of the laser ophthalmoscope, are able to treat successfully certain types of skin malignancies. The treatment is experimental and there has not been enough evidence to date to determine its effectiveness.

The most extensively applied use of the laser today is in

industry. Many companies have been formed to manufacture lasers for such industrial purposes as automatically cutting and shaping tough metals, drilling holes in hard and brittle surfaces, and welding dissimilar materials. Laser bullets strike the material at temperatures of 18,000° C, hotter than the surface of the sun, to produce a variety of complex industrial parts faster and more efficiently than conventional methods.

Cost is still a limiting factor in all laser applications, for unless the laser offers a technical advantage over mechanical tools it may be too expensive. Some of the factors favoring the use of the laser are: (1) The laser can weld different kinds of material together better than any other device. For example, it can weld metal to glass in such a way that only the immediate area is affected by the weld. The surrounding surfaces are not damaged or endangered by the excessive heat, as they are in the case of conventional welding. (2) Electron-beam welding can produce undesirable electrical discharges which affect the electronic components that may be involved in the process. The high-frequency beam of the laser plus the instantaneous weld avoid this possibility. (3) When holes have to be drilled in hard or brittle metals, the laser does not risk breaking the material, since there is no physical contact between the laser and the material. The laser beam simply "vaporizes" the center of the metal to produce the hole. (4) Since the laser beam is coherent, its beam does not "splatter" and it can perform its task at any convenient distance from the target area.

Laser beams aimed at the landing site of the Surveyor 7 spacecraft are shown as two tiny white dots side by side on the dark side (left) of the moon. This picture was taken by the TV camera aboard the spacecraft January 20, 1968. The lasers transmitting the beams were located at Kitt Peak National Observatory, Tucson, Arizona (right dot), and Table Mountain, Wrightwood, California (left dot). An argon gas laser developed by Hughes Aircraft Company was used at the Table Mountain site. (*Courtesy Hughes Aircraft Company*)

There is no all-purpose industrial laser, so it is important that the right laser be chosen for the right job. Some tasks require the powerful bursts of the pulsed beam; others, such as in drilling shallow holes, demand a Q-switching capability. On the other hand, if the holes require deep drilling, a continuous wave laser might be more efficient.

Drilling holes in diamonds which will be used in industrial tools is easy for the laser. The hardness of the diamond makes other methods slow and expensive. The average time required to drill a hole in a diamond by conventional means is three days. But the laser can do it in ten minutes at a cost of a few cents.

Because it is unnecessary for the laser to make physical contact with the material on which it is working, it can cut and shape one substance which is completely enclosed in another without damaging the outside material.

The laser beam is so fast, controlled, and positionable that it can perform the most difficult welds. Complete electronic circuit boards which can pass through the eye of a needle are welded by the laser, a task which would be extremely tedious and error-ridden even with the electron-beam welder.

The carbon dioxide laser is probably one of the most efficient tools available for cutting and welding. It can produce thousands of watts of power with very little pumping. With a special oxygen attachment it can outperform the plasma arc welding method by 25 percent. The laser destroys a minimum of material during the cutting and does virtually no heat damage. Since the laser

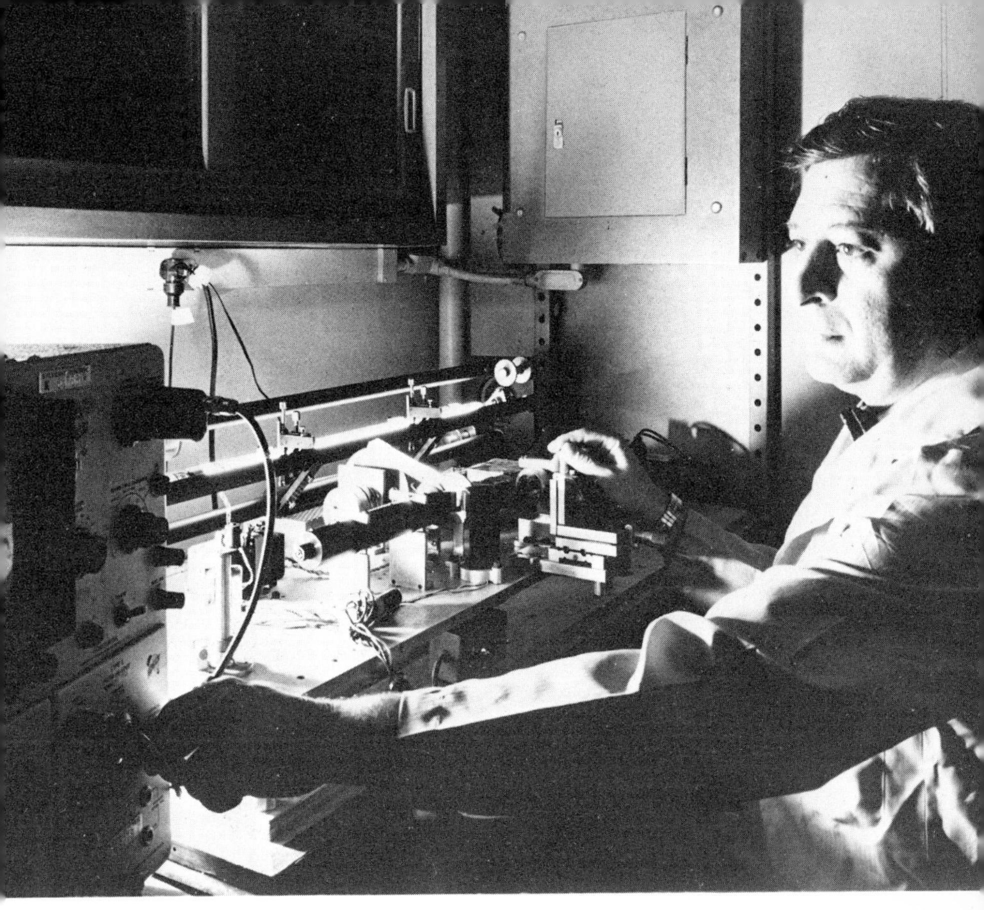

A laser microscope, which uses an infrared beam to look beneath solid surfaces, developed by General Telephone & Electronics Laboratories, Inc. (*Courtesy General Telephone & Electronics Corporation*)

beam is the only contact with the material, a need for lubricating liquids is eliminated, avoiding another common source of job contamination.

The use of the laser in engineering tasks is primarily in the areas of testing, measuring, guiding, and studying.

The laser beam provides a continuous and visible guideline for tunneling and mining machines to follow. The laser is usually attached to the wall of the tunnel, lined up for the desired path of the hole to be cut, then focused through the tunneling machine as a guide. The "mole" operator simply keeps the cross hairs on target and follows the laser beam. Laser guides have reduced the need for much advanced route surveying, saving weeks of construction time.

Lasers are finding increasing applications in wind tunnel tests to obtain air density patterns. They are used to monitor shock waves and rocket exhaust patterns. These measurements were extremely difficult to obtain before the advent of the laser, since the actions take place in millionths of a second under hazardous conditions.

As an important instrument in quality control, the laser enables the production of "perfect parts" with high repeatability by checking instantly such parameters as size, shape, and connection.

The laser has come a long way since its discovery in 1960—from a bulky, temperamental laboratory collection of mirrors, circuits, and flashing lamps with a tendency to overheat and in constant need of rest to a versatile power device, in some instances not much bigger than a pencil, with the reliability of a transistor radio.

6
Laser Safety

There are important safety rules to be observed around lasers. Although most laser areas are clearly marked with signs and warning lights, common sense and caution are still the best rules to follow.

The beam is the main source of laser danger. Because of its speed and concentration, it is possible to be injured by it before realizing it. Protective clothing is a good precaution in operating laser areas, and dark glasses should always be worn when the laser beam is on.

High voltage is another danger of which to be aware, since lasers store their power in capacitors and transformers which are always "hot," even when the laser is inoperative.

Coolant gases, such as nitrogen, should be handled with protective gloves and in well-ventilated areas.

Some lasers give off deadly fumes, such as carbon monoxide, so avoid breathing these exhausts.

Technicians wearing protective glasses in the vicinity of the argon gas laser, developed by Bell Telephone Laboratories, for use as a medical tool in "bloodless surgery." *(Courtesy, Bell Telephone Laboratories)*

Ruby laser system developed by Raytheon Company for drilling small holes in diamonds and other hard materials uses closed-circuit TV to enable the operator to safely position the diamond and monitor the drilling. *(Courtesy Raytheon Company)*

Research indicates that the seriousness of laser injuries depends on these factors: (1) the area of the body affected—eyes are the worst, (2) the power of the beam, (3) the wavelength—the shorter the more dangerous, and (4) the length of the exposure to the beam.

7
Laser Future

The potential applications of the laser are as wide as your imagination will let them be. Here are some predictions for the laser's future from the laboratories of industry, many of which are in the prototype stage today:

It is predicted that lasers could produce instant automatic fingerprint identification on the spot. There could be laser-powered rockets that can considerably reduce space travel time. Scientists conceive of a laser that can harness the power of the hydrogen bomb to give man an untapped work source. A 10,000-watt laser is on the drawing board which, when focused and projected by a 200-inch reflecting mirror, would be visible in the very far reaches of space—95 quadrillion miles.

The laser's future may be compared to the future of navigation after the voyage of Columbus. It was not just a matter of traveling a lot farther but of changing the concept of the world.

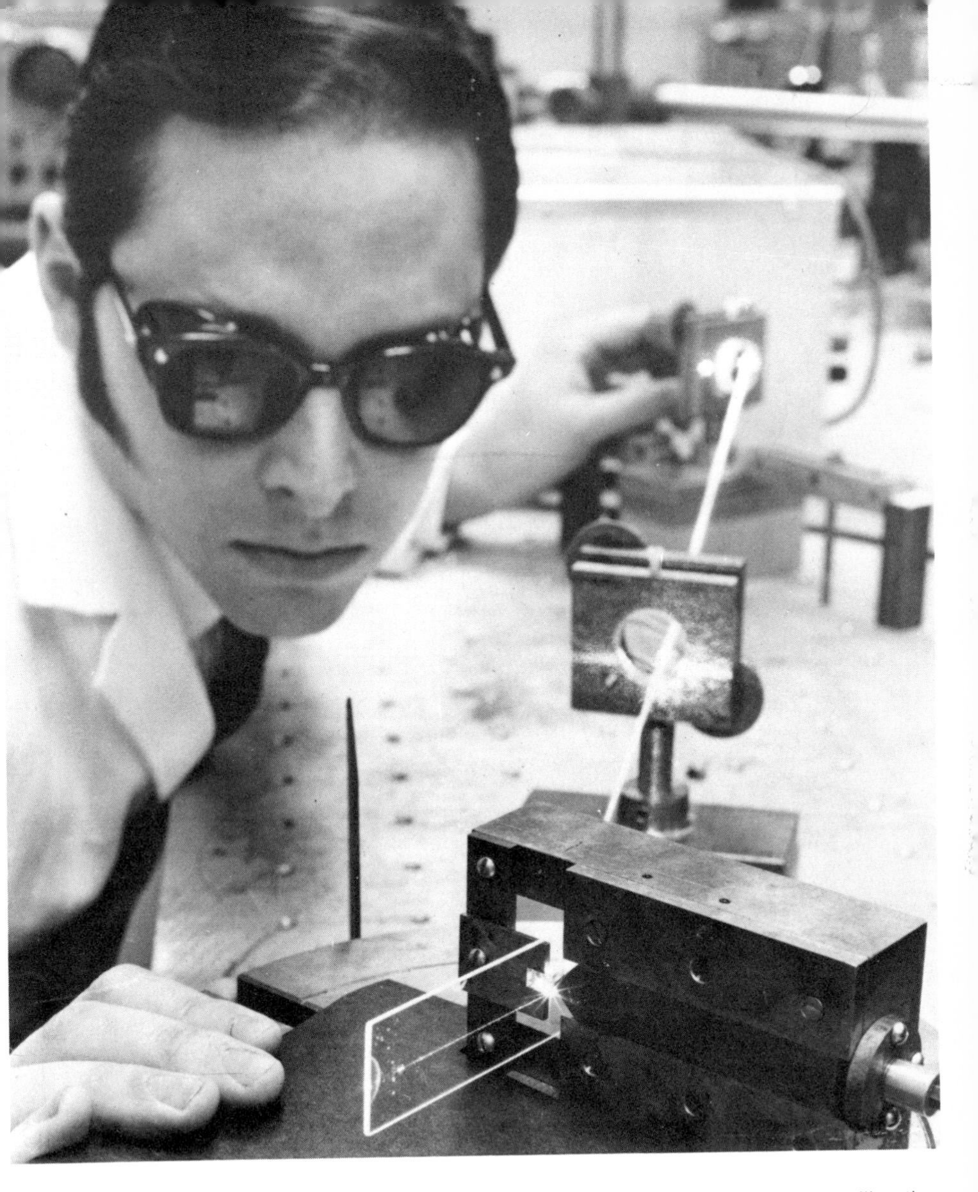

This glass plate has a deposit on the surface that makes it act as a light guide controlling the laser beam. Light guides can bend laser light and may be used someday as optical circuits in laser communication systems. (*Courtesy Bell Telephone Laboratories*)

The much shorter radiation wavelengths in the photoelectric region are making possible more accurate ranging and a promise of relief to our crowded communication channels. (*Courtesy RCA*)

Everything in our lives will ultimately feel the impact of the laser on it, from the most glamorous to the most routine—from a ray of light which will propel us through space to a weed killer.

And the laser itself—what of it? It will become smaller, yet more powerful, efficient, and safe. One writer describes the laser of the future as projecting X rays, not light. The X-ray laser will be so simple and efficient that all that will be necessary to change an application will be to insert a different lasing liquid in a tube.

And other lasers, higher up the spectrum, will emit gamma rays—the miracle workers—which will heal wounds, mend broken bones, eliminate disease, and restore sight.

By now everyone should be able to list his own predictions for the laser, and chances are they all will be correct. But before we range too far into this laser tomorrowland, let us look again at the laser today. Although in many tasks it still has to compete with conventional tools, technology is gradually proving the laser's superiority and society is making it a household word.

An experimental multitube CO_2 laser developed by the Raytheon Company. The purpose of the forty-foot-long tubes connected optically in series is to make possible large laser volumes in small space. The system was reported to have an output of 8,800 watts at 10.6 microns (infrared) and an efficiency of 15 percent. The two-inch-diameter tubes are water cooled. (*Courtesy Raytheon Company*)

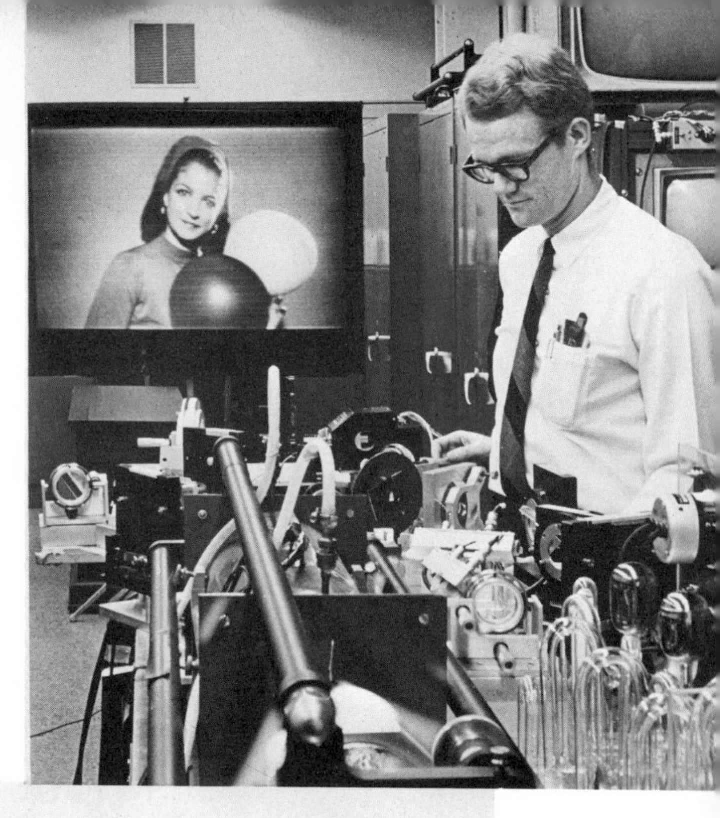

ON
EXT
POWER
SET

Above: A high-power CW argon ion laser developed by Hughes Aircraft Company for future applications in space communication, laser telephony, holography, and optical computer reading. *(Courtesy Hughes Aircraft Company)*

Top Left: Laboratory model of a laser display color TV system developed by General Telephone & Electronics Laboratories, Inc. *(Courtesy General Telephone & Electronics Corporation)*

Bottom Left: Talking and listening via laser beam is possible with this laser communicator developed by Hughes Aircraft Company. It can operate up to a distance of six miles — someday, hundreds of miles. *(Courtesy Hughes Aircraft Company)*

Glossary of Laser Terms

Beam diameter—The diameter of the laser beam, measured in millimeters.

Beam divergence—The spread of the laser beam, measured in milliradians.

Coherence—The property of staying in phase; the peaks and valleys of the wave energy occur at the same time and point in space.

Coherence length—The distance within which the waves stay in phase.

Diffraction—The bending of light waves as they pass the edge of a solid body or go through it.

Diffraction pattern—The alternating bright and dark dots or bands which make up the interference pattern.

Hologram—An interference pattern stored on film, formed by interference between a single direct wave from a laser and a reflected wave from an object. Generally viewed by illuminating it with a laser. It has all the 3-D properties of the real image, stores enormous amounts of information, and can be used as a memory.

Interference—The points of maximum wave energy do not occur at the same point in space because the waves are out of phase. This can be constructive when the energy of both waves combine, or it can be destructive when the energy of one is subtracted from the other.

Line selection—The use of a prism or filter in the optical cavity so that only one wavelength at a time can be selected by the laser. Adjusting the position of the prism tunes the laser to another wavelength.

Mode locked—A laser with an unusually stable pattern of longitudinal modes; required for holography, which takes long time periods.

Modes—Paths and patterns which light can follow within the laser and leaving it.

Multiline operation—Output at several wavelengths simultaneously.

Photon—A very small quantity of energy with properties of both a wave and a particle. All radiant energy (heat, light, X rays) comes in photons.

Power—Light output measured in watts or joules.

Power density—Power per unit of laser beam area, W/mc^2. It can be high, because of the small beam diameter and area. When focused to a point by a lens or mirror, the power density rises to very high levels due to the amount of power packed in a small area. It can then do damage with a milliwatt of power.

Quantum—The amount of energy added to or subtracted from an atom when a photon is absorbed or given off. Each atom will accept or release only certain amounts (quanta) of energy. Each of these corresponds to a photon of particular wavelength. Thus an atom of one material will only produce light of certain definite colors. No other material will have the same (color) fingerprints.

Stability—Generally used to describe the warm-up time of a laser and the time variation in power output thereafter.

Index

The Author

Charles H. Wacker, Jr., has been an industrial writer for several major American corporations. A graduate of Columbia University and the holder of a doctorate degree from UCLA, he has specialized in interpreting a variety of scientific subjects for the layman. He, his wife, and two children live in Santa Monica, California.